Robert William Parker

Congenital Club-Foot, its Nature and Treatment

With Special Reference to the Subcutaneous Division of the Tarsal

Ligaments

Robert William Parker

Congenital Club-Foot, its Nature and Treatment
With Special Reference to the Subcutaneous Division of the Tarsal Ligaments

ISBN/EAN: 9783337229955

Printed in Europe, USA, Canada, Australia, Japan

Cover: Foto ©berggeist007 / pixelio.de

More available books at **www.hansebooks.com**

CONGENITAL CLUB-FOOT:

ITS NATURE AND TREATMENT.

WITH ESPECIAL REFERENCE TO THE SUBCUTANEOUS DIVISION OF THE TARSAL LIGAMENTS.

BY

ROBERT WILLIAM PARKER,

SURGEON TO THE EAST LONDON HOSPITAL FOR CHILDREN
AND TO THE
GROSVENOR HOSPITAL FOR WOMEN AND CHILDREN.

...... " L'homme de l'art tire sa règle de conduite du cas qu'il a sous les yeux."......
DE SAINT-GERMAIN.

LONDON:
H. K. LEWIS, 136, GOWER STREET, W.C.
1887.

TO

WILLIAM ADAMS, F.R.C.S.Eng.

AS A SMALL ACKNOWLEDGMENT OF THE SERVICES HE HAS
RENDERED TO

ORTHOPÆDIC SURGERY

THIS

𝔐onograph is 𝔍nscribed

BY

THE AUTHOR.

PREFACE.

For some years past, I have been studying the subject of Club-foot, especially its mode of causation, and the anatomical condition of the structures chiefly concerned in the deformity. I am proud to acknowledge the valuable co-operation of my friend Mr. Shattock in this part of my work.

In the following pages, an attempt will be made to show :—

That certain of the tarsal ligaments are relatively short, owing to developmental malposition of the bones, between which they lie.

That this anatomical condition is CONSTANT, and is the chief impediment to the reduction of the deformity.

It will be argued therefrom :—

That simple tenotomy leaves the major part of the anatomical condition untouched.

That the relapses, which so frequently occur after tenotomy, and even after the prolonged use of mechanical appliances, are largely due to the unyielding nature of these ligaments ; and

That the remedy is to be sought in DIVISION OF THESE STRUCTURES.

THE AUTHOR.

LONDON ;
March, 1887.

CONTENTS.

	PAGES
DEDICATION	v
PREFACE	vii

INTRODUCTION.

Stromeyer and Subcutaneous Tenotomy—Little—Bouvier—His Tenotomy Knives and Modified Operation—Early Views as to Causation of Talipes—M'Keever—His Operation of dividing Ligaments—Streckeisen and his views—Gradual Rectification after Tenotomy—Relapses—Tarsectomy . . . . . . 1—7

CHAPTER I.

PATHOLOGICAL ANATOMY OF CONGENITAL CLUB-FOOT.

Dissection of Double Equino-Varus from a child, aged 18 months: Histological integrity of Nerve-Centres, of Nerve Trunks, and of Muscles. Condition of Tarsal Ligaments—Other Dissections—A shortened condition of the Ligaments is constant—The Astragalus in Man and in the higher Apes contrasted and compared . . 8—20

CHAPTER II.

ETIOLOGY OF CONGENITAL CLUB-FOOT.

(1) NERVE CAUSES:—Critical Analysis of the "Nerve" Causes—Not founded on Pathological Observation—Direct Observation to the contrary—Clinical Evidence to the contrary—A possible Factor in exceptional cases 21—25

(2) BONE CAUSES:—Critical Analysis of the "Bone" Causes—The undoubted occurrence of deviation in form—Evidence that the Bones were once normal—Deviations regarded as the consequence and not the cause—Analogous deviations in other Joints . . . 25—31

(3) MECHANICAL CAUSES:—Their sufficiency to explain majority of cases of Club-foot—Subdivision of these Agencies into Groups—Cruveilhier's Theory. Martin's Views, as to absence of Liquor Amnii—Author's Views—Examples of each Group of Causes—Club-hand—Heredity 31—52

CHAPTER III.

VARIETIES OF CONGENITAL CLUB-FOOT.

Anatomical Classification. Talipes Calcaneus and Calcaneo-valgus—Talipes Equinus and Equino-varus—Irregular Forms—Crus Varum—Anatomy of each variety: Ligaments—Muscles—Bones—Skin—Untreated and Relapsed Forms 53—70

CHAPTER IV.

TREATMENT OF CONGENITAL CLUB-FOOT.

(1) The Subcutaneous Division of Tendons and Ligaments—Objects of Tenotomy and of Syndesmotomy—Indications—Process of Repair in Tendons and in Ligaments—Immediate or gradual rectification after Operation—Rules for Division of Tendons—Rules for Division of Ligaments 71—84

(2) Treatment of Club-foot.
   (i) General Remarks on Treatment—At what age to commence—Manipulation—Galvanism—Flexible Splints—Scarpa Shoes—Plaster of Paris—Accidents—Tenotomy should be Subcutaneous 85—92
   (ii) Treatment of Special Forms. Talipes Calcaneus and Calcaneo-valgus—Talipes Equinus and Equino-varus—Irregular Forms—After-treatment of Talipes—Prognosis—Tarsectomy—Conclusion 92—102

INDEX 103

THE NATURE AND TREATMENT

OF

CONGENITAL CLUB-FOOT.

INTRODUCTION.

To STROMEYER belongs the honour of having introduced subcutaneous division of tendons into surgical practice, and thus indirectly, the credit of having founded on its present basis the principles and practice of orthopædic surgery.

The discovery was introduced into this country, and first practised by one of STROMEYER's earliest patients—Dr. LITTLE—to whose courage and subsequent practice and teaching subcutaneous tenotomy owes much of its present success. It was largely Dr. LITTLE's enthusiasm which helped the new operation over its earliest difficulties and over the opposition it at first met with on the Continent, and in this country also.

It is not a little remarkable that the treatment of club-foot by subcutaneous tenotomy should still be carried out almost exactly as STROMEYER and LITTLE first practised it, and that no important modification should have been made or attempted on the original plan of operating, although fifty years have elapsed since STROMEYER's* first

* 'Rust's Magazine,' Bd. 39, S. 195, 1833.

operation, and notwithstanding the great activity which has characterised orthopædic surgery during this period.

It is true that BOUVIER* improved on STROMEYER's operation; he used smaller knives, and only made one puncture. The sharp-pointed bistouri was simply used to incise the skin; then a blunt-pointed tenotome was passed in beneath the tendon, which was cut in the way now usual with us. Such a tenotome, he thought, was less likely to injure vessels or nerves, and there was less fear in dividing the tendo Achillis, of the skin being wounded at the point opposite to its entrance.

And not only as regards operative treatment has the subject of talipes remained more or less stationary, but the pathology and etiology of the deformity have likewise remained where STROMEYER left them; his views, since emphasised and upheld by Little, still predominate, a remarkable instance of the "influence of authority in matters of opinion."

Dr. LITTLE,† says that club-foot results from "any cause, whether paralytic or spasmodic, by which the equilibrium between different sets of muscles that are naturally antagonistic is disturbed, produces the distortion vulgarly called club-foot," p. 23. The general acceptance of this view has doubtless largely contributed to draw chief attention to the muscles. In the earlier periods of tenotomy, moreover, the cases treated were mostly adults, in whom secondary deformities played a large share, and in whom the muscles would naturally be largely implicated. The recorded dissections also are on feet long past the

* 'Mémoire sur la Section du Tendon d'Achille dans le traitement des Pieds-Bots,' Paris, 1839.

† 'A Treatise on the Nature of Club-foot and Analogous Distortions,' London, 1839. This work contains an admirable summary of opinions entertained respecting the morbid anatomy of club-foot, "from the earliest writers to date of publication."

period of infancy; many, indeed, are on adult feet in which the secondary deformities outweighed the primary. On the whole, therefore, it is not surprising that the nerve theory continued to prevail, or that it should direct surgical attention almost exclusively to those structures which the nerves were thought capable of influencing.

STROMEYER* refers to the permanent tendency of the feet to turn inwards after tenotomy, and says, " One can hardly doubt that this abiding tendency depends on the condition of the ligaments, which are shortened on the inner side of the foot and lengthened on the outer side. Such a foot is like a door which is not hung straight. Whatever force be used the door will not open straight. As the shortened ligaments cannot be reached with the knife, there remains nothing else for this condition but the application of apparatus . . . ."

STROMEYER, at this early period of subcutaneous tenotomy, can hardly be reproached for not having attempted section of these shortened ligaments, for even scubcutaneous tenotomy was itself only introduced into general surgical practice after much opposition from men who were regarded as leaders of surgical opinion at that period. Moreover, the results obtained from tenotomy in many cases were justly regarded as so brilliant compared with what had been previously obtained from purely mechanical measures, that a suggestion to cut the tarsal ligaments at that time would very probably have endangered the future of the milder measure. BOUVIER (op. cit.) says, " A few days suffice to replace talipes equinus in the natural position, when there is no resistance in the ligaments or in the bones." These authors, however, had long been anticipated with regard to the part played

* 'Beiträge zur Operationen Orthopaedik,' Hannover, 1838.

by the ligaments in maintaining the deformity. Dr. Thomas M'Keever, in a paper* on club-foot, dated Dublin, November 2nd, 1819, says : " The lower extremity of the tibia rested on the os naviculare, the apposed surfaces being smooth, though not so as in an ordinary joint. The internal cuneiform and navicular bones were attached by means of strong ligamentous bands to the inner ankle. I was now proceeding to examine what share each individual bone had in producing the deformity, and for this purpose divided the ligamentary band just spoken of, when I was much surprised to find that I could, with the greatest facility, reduce the foot into its natural shape."

Many subsequent writers, including GUÉRIN, LITTLE, LONSDALE, ADAMS, have referred, more or less at length, to the part played by the ligaments, but I am not aware that any of them has advocated their division in suitable cases as routine treatment. In the year 1868, however, a posthumous paper by Professor STRECKEISEN, of Bâle, on club-foot was published,† in which as the result of careful dissections, the author not only recognised the part played by the ligaments, but in which he recorded cases successfully treated by their division, and boldly advocated the practice.

After section of the tendons, STROMEYER, as DELPECH‡ and others before him, taught that the foot should be retained in its old position until union of the divided ends had taken place; he feared that, if separated, union between the divided extremities might not take place, or that inflammation might be set up. THILENIUS,§ who cut

* 'Edinburgh Medical Journal,' 1820, vol. xvi, p. 220.
† "Notizen über Bau und Behandlung des Klumpfusses." 'Jahrbuch für Kinderheilkunde.' Band 2, 1869, p. 49, et seq.
‡ 'De l'Orthomorphie,' Paris, 1828.
§ 'Medecin. und Chirurg. Bemerkungen,' Frankfurt, 1789.

the tendo Achillis by the open method, had boldly put the foot into its normal position, a complete though slow cure of the deformity resulting. BOUVIER recommended that extension of the foot should be commenced almost at once.

At the present day a similar difference of opinion exists as to the advisability of gradual extension or the immediate reposition of the deformed foot. The weight of authority is still on the side of gradual extension commenced two or three or more days after operation, although although there is no lack of evidence to demonstrate that union takes place just as certainly if the other method of procedure be adopted, that considerable time is gained and some pain saved the patient.

Surgeons generally recommend an interval of some days at least between section of the tibial muscles and of the tendo Achillis, the former preceding the latter. On this point, too, there are differences of opinion, and the more rational practice of removing the greater impediment first or simultaneously with the lesser appears to be gaining ground. There can be little doubt that section of the tendo Achillis alone often suffices; the weaker muscles yielding readily to mechanical treatment. STROMEYER seldom found it necessary to divide the tibial tendons, contenting himself, in the majority of his cases, with section of the Achilles tendon.

In the early days of subcutaneous tenotomy a large proportion of the cases of club-foot operated upon were long past the age of infancy. Many were adults or young adults, some were children. It is probable, however, that in the majority secondary conditions existed which complicated the treatment and rendered a cure much less easy of attainment. Under the circumstances it is not at all strange that relapses were frequently met with; rather should the pioneers in this branch of surgery be congratulated on the many successes they obtained. Now, as formerly, many cases relapse after apparently successful treatment; this is ascribed to a cessation of that care and supervision which a club-foot constantly demands in order to completely overcome its abiding tendency to reassume the position in which it was created, and to which from long use, or in obedience to still active causes, it has become naturalised. This explanation, however, is not entirely satisfying; there is something more than want of supervision, as I shall endeavour to show further on.

It may be asked whether any real advances have been made by modern surgeons in the treatment of the deformity. It is true that a great amount of mechanical skill has been lavished on the apparatus with which the treatment, subsequent to tenotomy, is carried out; but in other respects everything remains much where STROMEYER, BOUVIER, and DIEFFENBACH* left it years ago.

Some of these relapsed cases appear so hopeless that the most radical surgical measures are now regarded as justifiable and even called for. Thus, among many others, Dr. LITTLE proposed (and Mr. SOLLY carried out for him) excision of the cuboid bone. Mr. LUND has removed the astragalus. Mr. DAVIES-COLLEY has removed a wedge-shaped portion from the tarsal bones. Mr. PUGHE has resected the head of the astragalus. Mr. DAVY recently read a paper on the "Radical Cure of Club-foot" at the Royal Medical and Chirurgical Society,† in which he advo-

---

* 'Ueber die Durchschneidung der Sehnen und Muskeln,' Berlin, 1841.

† 'Medico-Chirurgical Transactions,' vol. lxviii, 1885, also the Society's 'Proceedings,' vol. i, p 338, with the discussion thereon.

cated removal of an osseous wedge from the transverse tarsal joint.

From the foregoing summary it will be seen how little the subject of club-foot has advanced, that on many points the same uncertainties exist now as formerly, and that much yet remains to be accomplished before the pathology and treatment of club-foot can be considered as finally settled.

I venture to think that early treatment directed against the structures chiefly at fault—the ligaments—will materially lessen the tendency to relapse as well as the number of cases in which tarsectomy need ever be entertained.

CHAPTER I.

PATHOLOGICAL ANATOMY OF CONGENITAL CLUB-FOOT.

Dissection of double equino-varus from a child aged eighteen months—Absolute integrity of spinal cord and nerves, and of the muscles—Condition of the tarsal ligaments —Other dissections—A shortening of the ligaments constant—Conformation of astragalus in man and in the higher apes contrasted and compared.

The opportunities for studying the anatomy of congenital club-foot in infancy and early childhood before secondary deformity has had time to develop, are, comparitively speaking, few and far between. The observations I have to offer are founded on the dissection of several cases of my own, on a study of the specimens in the Museums of the Royal College of Surgeons of England, of St. Bartholomew's, St. Thomas's, St. Mary's, University College and Guy's Hospitals; on a study of the conformation of the astragalus in man and the higher apes, and on cases of club-foot seen and treated by me at the East London Hospital for Children during the past ten years.

I have great pleasure in acknowledging indebtedness to my friend Mr. Shattock, for his great help in the preparation of this portion of my monograph; it is very largely founded upon a paper written jointly with him, and published in the thirty-fifth volume of the Pathological

Society's 'Transactions' (1884). I must not omit to thank him also for the drawings which illustrate the anatomy of the deformity. For their accuracy I can vouch; their artistic merit speaks for itself.

I will anticipate by just saying that, in dissecting these talipedic feet, one of the first, most salient, and most constant points to strike me was the condition of the ankle and tarsal ligaments. I found after removing all the muscles that the deformity remained unaltered, and that it could not be overcome without using much more force than would be tolerated by the living foot. Since adopting the plan of treatment advocated further on, my relapses have been fewer, and the time necessary for successfully treating cases of club-foot has been materially shortened.

I will now give the details of the dissections, and then proceed to discuss how they bear upon the generally accepted pathology of club-foot, and on its surgical treatment.

Dissections.

CASE 1.—The specimens were removed from the body of a child aged 18 months who died in the East London Hospital for Children of tubercular meningitis. I first saw the child as an out-patient when five weeks old; she was the subject of well-marked double congenital equino-varus. Treatment by manipulation and plaster bandages was commenced at once, and some progress made; the child, however, ceased to attend, and when seen again fourteen months subsequently, at the commencement of her fatal illness (tubercular meningitis), the feet were as bad as ever. A complete autopsy was made; but reference will only be made in this place to the spinal cord and the affected limbs.

The right foot was dissected so as to display the muscles and tendons *in situ*; the ligaments and bones of the other foot were specially studied, portions of each of the muscles and of the chief nerve-trunks being examined microscopically.

The spinal cord presented no naked-eye change either externally or on section; there was no alteration in its consistence at any part, nor any change in its membranes, or in the spinal canal. It was hardened in a weak solution of chromic acid frequently changed; the sections were stained with hæmatoxylin and with carmine, and with the two combined. They were mounted, some in dammar, some in Canada balsam. Many sections from each of the regions were examined under the microscope, but nothing abnormal could be seen anywhere. I may specially remark that the multipolar cells in the anterior horns of the grey matter were quite healthy in appearance, and normal as to number. The popliteal nerve and its main divisions, prepared in the same manner as the cord, were also examined with a like negative result.

Portions of each individual muscle of the left leg, after hardening in spirit, were examined. The sections were stained, some with osmic acid, some with hæmatoxylin, and found to be perfectly healthy.

After removing all the muscles from the left leg I was struck by the fact that considerable force was still required to straighten the foot. On removing the anterior portion of the internal lateral ligament of the ankle, which was found to be firmly blended with the astragaloscaphoid ligament above, and with the calcaneoscaphoid ligament below—both of which ligaments were considerably shortened—this resistance was overcome. A bursa was found between the tip of the malleolus and the navicular bone.

The astragalus presented some remarkable deviations from the normal. Its trochlear surface was extended backwards as far as the posterior edge of the lower articular surface. The extent of this additional surface was still easily

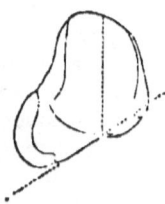

FIG. 1.—Astragalus, from a child aged 18 months, with equino-varus. Obliquity of neck = 53°. Natural size.

recognisable; and it was clearly due to the condition of extreme extension of the ankle joint. On the contrary, the extent of the trochlear surface was proportionately lessened in front, owing to the fact that it no longer formed part of the proper articulating surface. In the next place, the neck of the astragalus was lengthened and directed inwards with an unnatural obliquity; measured in the manner about to be described, this obliquity amounted to 53° (Fig. 1), as against the mean 49·6° in varus cases, and 38° in the healthy bone. The articular surface of the head was prolonged on its inner side; and instead of being uniformly convex it was divided into two parts, the planes of which met at an obtuse angle; the inner and larger corresponded to the displaced navicular bone; the outer portion, which looked forwards, was unopposed, instead of being, as in the normal condition, in contact with the navicular. The redundant portions, both on the head and on the trochlear surface, corresponded in extent with the normal limits, but did not present the polished surface of the rest of the articulating areas, being covered with a layer of loose connective tissue. The internal malleolar facet was unrecognisable, doubtless because it, too, had ceased to form part of the proper articulating area.

The calcaneum lay in a position of exaggerated rotation inwards beneath the astragalus, a considerable portion of the upper posterior facet being uncovered and marked off from the rest by a low ridge, similar to that found on the head of the astragalus and above referred to. In consequence of the extreme extension of the ankle-joint, this uncovered portion of the posterior facet articulated with the posterior border of the external malleolus. The inner portion of the posterior facet was continued into that on the sustentaculum. The plane of the cuboidal facet was directed unnaturally inwards, and its outer border less prominent than usual. These results were due to a curving inwards of the anterior part of the bone from traction made upon it through the external calcaneo-cuboid ligament. It may also be mentioned that the cartilage-basis of this bone was structurally continuous with the navicular, a condition alluded to by CRUVEILHIER in his 'Anatomie Pathologique.' The condition is comparable to that in which the digital phalanges are sometimes coalesced or connate.

CASE 2.—The left foot from an anencephalous fœtus at about the seventh month. I would especially observe that the inversion of the foot could not be completely overcome until all the ligaments passing between the several bones on the inner side of the foot had been divided. The foot was in a position of well-marked equinovarus. No deviation in the form of the astragalus could be discovered, except that the angle of obliquity was considerably less than normal, being only 31° as against 38° the mean

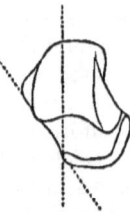

Fig. 2.—Astragalus, from a 7 months' fœtus, with one-sided equino-varus. Obliquity of neck unusually small = 31°. Natural size.

in health, and 49·6° in equino-varus cases. The calcaneum was normal. The right foot was normal.

Case 3.—From an anencephalus fœtus of about the seventh month. Left foot in a position of extreme equino-varus. Before dissection the inner border of the foot was capable of being brought up to the inner side of the leg; extension of the ankle beyond a right angle was impossible. On dissection the upper extremities of the tibia and fibula were found in their normal relative position to one another and to the femur, whilst the lower end of the fibula lay in a plane anterior to the internal malleolus, the transverse axis of the lower end of the tibia being directed forwards and outwards (a similar condition has been described by Jörg). It was evident, therefore, that the lower ends of the tibia and fibula had undergone a marked amount of rotation inwards. The internal malleolus was in contact with the navicular, and the navicular with the sustentaculum tali, a bursa intervening. The body of the astragalus maintained its proper position relatively to the tibia and fibula; the bone was in a position of full extension. The

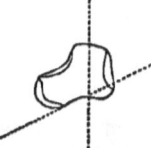

Fig. 3.—Astragalus, from a fœtus with extreme equino-varus. Obliquity of neck = 64°. Natural size.

cartilage of the trochlea in front of the surfaces, actually in contact, was impressed by the fasciculi of the anterior ligament.

ADHESIONS WITHIN ANKLE-JOINT.

Band-like adhesions existed between the trochlear and the opposed margins of the tibia and fibula. The navicular facet was divided into two, the respective planes being at rather less than a right angle to one another; the outer division was of small extent, and covered by the capsule of the joint; the inner one alone was in contact with the navicular. The obliquity of the neck in this case was extreme; as measured in the manner hereafter described it represented an angle of 64° as against the mean 49·6° (Fig. 3). On placing the calcaneum in an antero-posterior plane, its cuboidal facet lay at an angle of 45° with this plane, and the outer margin of the bone was rounded as described in Case 1.

CASE 4.—From an undersized fœtus of about eight months, with talipes calcaneus of the right foot. The ankle did not admit of extension to a right angle. On dividing the tendons of the tibialis anticus and the extensor proprius hallucis extension-movement was increased, but division of the anterior ligament of the ankle was necessary before full extension was possible. Movements, especially of extension, in front of the transverse joint were also much limited. The outer and inner borders of the foot were convex

FIG. 4.—Astragalus, from a fœtus with calcaneus. Obliquity of neck = 33°. Natural size.

towards the plantar aspect in all positions allowed to the foot, the summit of the inner curve corresponding with the junction of the first cuneiform and first metatarsal bones. The astragalus presented no recognisable deviation from the normal form, except that its fibular surface was vertically ridged down the middle, the anterior portion only corresponding with the malleolus in the fully flexed position of the foot. The angle of obliquity of the neck measured 33° (Fig. 4). The calcaneum was quite normal in all respects.

CASE 5.—Removed from a child, born at seven and a half months, recently under my care in the Children's Hospital with imperforate rectum, death taking place on the eighteenth day after left lumbar colotomy. There was talipes calcaneus of the right foot, the left foot being but slightly affected. After dissecting away the muscles, the anterior ligament was found to prevent extension-movement at the ankle-joint. The trochlear surface of the astragalus in this case was prolonged forwards on the upper surface of the neck of the bone,

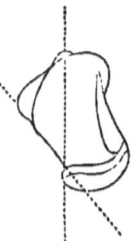

FIG. 5.—Astragalus, from a 7½ months' child, with calcaneus, which died on the eighteenth day after birth. Obliquity of neck = 39°. Natural size.

as far as the margin of the navicular facet. The inner malleolar facet was also prolonged on the inner side of the neck, nearly as far forwards as the limit of the margin of the navicular surface. The angle of obliquity of the neck measured 39° (Fig. 5). The calcaneum presented nothing abnormal.

CASE 6.—Fœtus at full term, with a sloughing spina bifida in lumbo-sacral region, and extreme equino-varus of both feet. On the left foot there was a depressed, atrophied patch of integument over the external malleolus, and a second one further forwards over the prominent outer portion of the head of the astragalus; a similar condition existed on the right foot. On raising the skin in this situation, a distinct lenticular bursal cavity, about 7 mm. in diameter, was exposed, containing glairy fluid (a condition similar to that described by VOLKMANN and LÜCKE). The plane of the facet on the astragalus corresponding to the navicular looked directly inwards; the external portion of this facet was well represented, the planes of the two divisions meeting at a right angle; but the portion of the joint corresponding with it, though persistent, was separated from the rest by a delicate line of adhesion. A well-marked band of adhesion passed from the trochlea, at its line of union with the fibular facet, to the tibio-fibular junction from front to back. The obliquity of the neck amounted to 56° Between the front of the trochlea and the capsule of the ankle-joint was a circumscribed cavity, separated by adhesions from the actual articulating area. The outer margin of the cuboidal facet of the calcaneum was rounded, as described in other specimens. For the opportunity of dissecting this specimen I have to thank Mr. D'Arcy Power, Curator of the Museum, St. Bartholomew's Hospital.

Many of the foregoing specimens have been presented to, and are to be seen in, the Museum of the Royal College of Surgeons.

I will next proceed to consider the normal conformation of the astragalus in the human adult and fœtus, as well as, for purposes of comparison, in the anthropoid apes. Attention has chiefly been directed to the extent and direction of the articular facets and to the obliquity of the neck of the astragalus.

To determine precisely the degree of this obliquity, the following plan of measurement was adopted: the astragalus, with its trochlear surface upwards and horizontal, was placed beneath a fine thread fixed across it; a second thread was fixed at right angles to this along the mid-line of the trochlear surface, parallel with its inner border; whilst a third was placed along the outer margin of the neck of the bone, so as to intersect the other two; the angle formed by the meeting of the two threads last described was taken as the measure of the obliquity of the neck. The subjoined woodcuts, from which, however, the transverse lines have been omitted, will further explain this.

In this manner the obliquity of the neck in twenty specimens of adult astragali, taken promiscuously, was measured. The mean angle was found to be 10·65°; the maximum was 26°; in three cases the angle was so small that to measure it was impracticable. In two cases only the angle amounted to 20° and upwards; in twelve cases it did not exceed 12°, and in the majority was under 10°. In the fœtus, from about the fourth month up to term, the mean angle in eleven cases was 38°, the maximum 42°, and the minimum 35°. In three cases only did the angle exceed 40°. In two cases of calcaneus this obliquity amounted to 33° and 39° respectively, being an average of 36°. In five cases of equino-varus the mean angle was 49·6°, the maximum 64°, the minimum 31° (from Case 2), being an average of 49·6°.

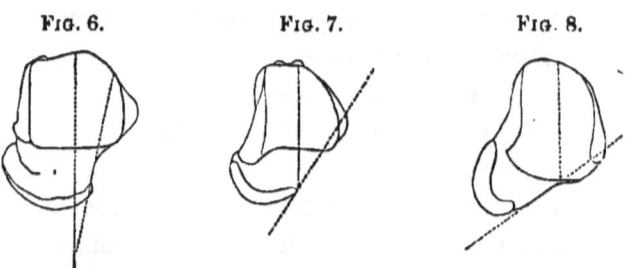

FIG. 6.   FIG. 7.   FIG. 8.

FIG. 6.—Astragalus (normal), from an adult. Obliquity of neck = 12°. Reduced.

FIG. 7.—Astragalus (normal), from fœtus at term. Obliquity of neck = 35°. Natural size.

FIG. 8.—Talipedic astragalus. Obliquity of neck = 53°. The sagittal line, if prolonged, would fall outside the navicular facet. Natural size.

By comparing Figs. 6 and 7, the difference between the adult and the infantile astragalus will be seen; whereas a line, drawn sagittally over the middle of the trochlear parallel with its inner border, crosses the navicular facet of the astragalus in the adult, in the child this same line lies outside the facet. In Fig. 8, that of a talipedic astragalus from Case 1, the sagittal line lies considerably outside the navicular facet, and the contrast between this figure and Fig. 6 is still more marked. From the foregoing measurements it is evident that an important difference exists at these periods of life; I submit that the modified form of the fœtal astragalus is to be associated with the high capacity for inversion enjoyed by the fœtal foot.

A comparative study of the astragalus discloses, as a normal condition, a corresponding obliquity in the anthropomorpha, in which animals the movement of inversion in the pes is one of the most ready of all the movements of the member; in association with the opposable hallux, the object of this movement of inversion for purposes of arboreal progression is obvious. The end gained by this obliquity is clearly the increased range of the movement of adduction associated with inversion of the foot, in which the navicular is carried round to the inner side of the head of the astragalus, a movement which would be hampered or prevented were the inner border of the neck to lie in a sagittal plane with the body of the bone.

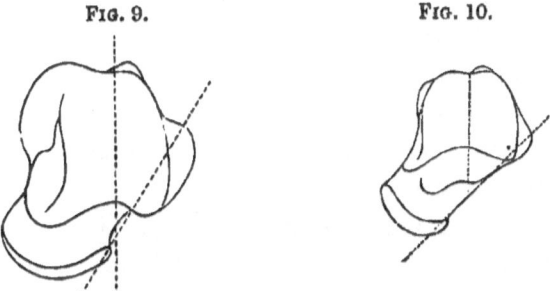

Fig. 9.—Astragalus, from adult chimpanzee. Obliquity of neck = 27°. The sagittal line falls outside the navicular facet. Reduced.

Fig. 10.—Astragalus, from young adult ourang. Obliquity of neck = 45°. The sagittal line falls at some distance outside the navicular facet. Natural size.

In the anthropomorpha it will be observed also that the sagittal line lies altogether to the outer side of the navicular facet. The difference in this regard between the simian, the fœtal, and the adult human astragalus will be plain from a study of Figs. 9 and 10; the approximation of the human, foetal, and especially of the talipedic astragalus, to the simian type will also be recognised.

I would draw especial attention to the prolongation of the internal malleolar facet on the inner side and upper surface of the neck of the fœtal astragalus, an anatomical character not hitherto described. This prolongation is, almost without exception, to be found in the fœtal astragalus; sometimes it is so marked as to approach closely to the inner border of the navicular facet. Its presence is an interesting fact taken with its persistent condition in the apes, in which animals the same extension of this facet exists. The explanation of its development may probably be found in the obliquity of the neck of the astragalus, and in the flexed position of the foot during the later months of intra-uterine life, a position which, in different degrees, is maintained for a varying period after birth. In this position the facet will be found in accurate contact with the internal malleolus. In the great majority of adult bones this anterior prolongation has disappeared, a fact to be associated with the diminution in the obliquity of the neck of the astragalus as age advances, and the perfect assumption of the upright position.

In the adult ape, however, the facet persists, and with it the obliquity of the neck and the high capacity for inversion. The persistence of this facet in the anthropomorpha is explained by the permanent obliquity of the neck of the astragalus and by the fact that these animals are unable to maintain the upright position, and in walking or standing keep the joints of the lower limb in a greater or less degree of flexion. In the human carpus, the presence of an *os centrale* has been demonstrated at an early period of fœtal life, such as forms a persistent carpal element among Simiæ and Rodentia. With such a passing character in the human carpus, this transitory conformation of the astragalus may, perhaps, be compared. In exceptional cases, in which the assumption of the upright position is delayed (as in some cases of rickety incurvation of the tibiæ), I have observed the capacity for inversion of the foot to persist up to the third or fourth year of life.

From the foregoing evidence it is clear that a capacity for inversion may be considered as physiological during fœtal life, and is a position which the foot, from the conformation of the bones, most naturally assumes. DIEFFENBACH, many years ago wrote (op. cit., p. 83) :—
". . . . All small children have a decided tendency to club-foot. . . . . The child retains this as long as it is carried in arms and only gradually loses it on walking, when the weight of the body presses the soles of the feet flat to the ground."

It will be conceded that the abnormal condition of the articular surfaces, and the altered position of the bones themselves, tend to place the muscles at a great disadvantage. The mechanism of the foot is such, that deviations from the normal reduce the power of the muscles, the latter being powerful largely because of the mechanical advantages they enjoy in the healthy foot; these advantages once destroyed by deformity, the muscles become comparatively powerless; and, it will be remembered, that they are invariably atrophied in neglected cases.

The considerable proportion of cases in which one foot only is affected, is a strong argument against the supposition that the deformity is due to an inherent want of right development, and the condition of the articular facets on the astragalus, to be presently described, further negative such a view. On the other hand the fact may be readily understood, if a mechanical mode of causation be invoked;

the mode in which mere position of the feet may lead to the production of varus on one side is, I think, shown by such a specimen as that represented in Fig. 11, a healthy fœtus of about twelve weeks within its normal membranes

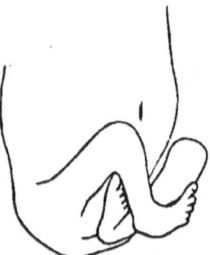

FIG. 11.—Suggested mode of production of one-sided deformity.

(St. Mary's Hospital Museum, North Collection, No. 11). Here it will be seen that the crossing or free foot retains the physiological position of inversion, while the left foot in the act of passive crossing has met the inner side of the opposite thigh and lies protected in the flexure of the knee. In such a case it is obvious that compression would maintain or exaggerate the inverted position of the crossing foot, whilst its fellow would be guarded from pressure owing to its position. The possible varieties of position, however, which would lead to a like result are, of course, numerous.

CHAPTER II.

THE ETIOLOGY OF CLUB-FOOT.

(1) *Critical analysis of the "Nerve" causes; not founded on pathological observations—Direct observations to the contrary—Clinical evidence to the contrary—A possible factor in exceptional cases.*
(2) *Critical analysis of the " Bone " causes—The undoubted occurrence of deviation in form—Evidence that the bones were once normal—Deviations regarded as consequence and not cause—Analogous deviations in other joints.*
(3) *Mechanical causes—Their sufficiency to explain majority of cases—Subdivision of these agencies into groups—* CRUVEILHIER'S *theory—*MARTIN'S *views—*AUTHOR'S *views—Examples of each cause.*

BEFORE giving my own views on the causation of congenital talipes and the mode of its production I will endeavour to find out how far the foregoing anatomical details support, or otherwise, the doctrine so widely held, that (1) talipes depends on some abnormal condition of the nerve-centres, or (2) on some abnormal condition of the tarsal bones. I will subsequently give my reasons for believing that in the majority of cases the deformity depends on environment rather than on pathological causes proper, and endeavour to show that clinical facts not only suggest but bear out this conclusion. I would not deny in a few exceptional cases, the immediate agency of other causes, such as disorder or disease of nerve-centres, or pathological deviations affecting the bones. Unfortunately this admission brings me no tangible satisfaction : for the production of these abnormal conditions are themselves, not improbably, due to mechanical causes at present not very clearly explained.

1. NERVE CAUSES.—Dr. LITTLE, who is one of the chief exponents of this view and the author of the essay on talipes in our largest English text-book of surgery, writes (op. cit., p. 24), " Having thus offered my opinions of the causes of those deformities of the feet which take place after birth, and stated the identity of their symptoms and morbid anatomy with those of the club-foot with which children are born, the probability will, I think, appear obvious that the remote causes are the same; but there are other phenomena connected with the history of these affections which render the accuracy of these opinions almost capable of demonstration. Fœtuses which have suffered some evident derangement in the development of the nervous system, such as those denominated hemicephalous and acephalous, or affected with spina bifida, and those born before the expiration of the normal period of utero-gestation, are particularly obnoxious to this deformity of the feet. The occurrence of the perfectly analogous deformity of the hands which takes place prior to birth, denominated club-hand, in which the flexors and pronators (analogous to the so-called extensors and adductors of the foot) are likewise contracted, corroborates the opinion that congenital club-foot depends on spasmodic muscular contraction. In the instances which I have examined of congenital deformity of the hand (club-hand), both in museums and

in the living subject, the feet were also affected with talipes, proving the operation of a common cause. . . ."

Again, and much more recently, he says,* "A comparison of club-foot with the distortions which occur after birth, unmistakably from disease of the nervous system, tends to prove that congenital and non-congenital club-foot spring from analogous causes. Distortion after birth from altered innervation of muscles is more common in the lower extremities, and especially in the feet, than in any other part of the frame. Club-foot is also the most common distortion before birth. After birth talipes varus, in consequence of cerebro-spinal affection, is more common than valgus. After birth, foot deformity, from disease of the nervous system, attains oftener a higher grade on the left than on the right side. This is equally the case with congenital club-foot. Some other agency than accidental uterine or pelvic pressure is required to account for these analogies; they cannot be regarded as mere coincidences."

I feel bound to dissent from these views for many reasons. In the first place, the large majority of children born with talipes are well formed, and in all other respects healthy; moreover, the supposed nerve-lesion in such cases has never been demonstrated. Then, apart from the fact that in the first of the cases related (p. 10) the nerve-centres and the nerve-trunks as well as the muscles of the limbs were histologically normal, there are clinical facts which tell against the nerve-theory of causation. The most important of these is that talipes is an accidental, and not an essential, sequel of paralysis. VOLKMANN† has shown that the paralysed foot falls into a position of equino-varus by gravity; it is, furthermore, the position of perfect rest

* 'Holmes's System of Surgery,' 3rd edit., vol. ii, p. 232.
† 'Pitha and Billroth's Handbuch,' vol. ii, chap. 48.

which the foot assumes during deep sleep or during artificial anæsthesia; it is a position which the foot is very apt to assume after fracture, and, indeed, whenever the limb passes out of control of the will; moreover, the foot can very easily be prevented from assuming such a position. In cases of infantile paralysis I have noticed further that the foot, even when left to itself, does not always assume a talipedic condition, and when talipes does supervene there are other unmistakable signs of nerve lesion, viz. general atrophy of the limb, with lowering of temperature, conditions which are persistent throughout life.

If further evidence in this direction be needed it is to be found in the fact that congenital malformation of the nerve-centres occurs without the association of talipes or atrophy of limbs, as is abundantly shown by the specimens of an encephalus preserved in museums. I have seen cases of spina bifida without deformity of the feet, but with atrophy of the lower limbs, probably due to nerve change. I have also seen cases with atrophied and weakened limbs, and unilateral talipes, the opposite foot showing no tendency to assume a talipedic condition. In eleven consecutive cases of spina bifida which have been under my care, there was no talipes in seven. In one case there was marked double equino-varus; in two double, and in one single calcaneus; while in almost all the cases there was more or less motor paralysis of the limbs, with evidences of want of nutrition, sometimes alone, sometimes associated with paralysis or paresis of the sphincters.

But I think it quite possible in a few cases of talipes that there may be a nerve lesion apart from such manifest conditions as spina bifida, &c. For very occasionally at the time of birth (and the same may be found in the fœtus at comparatively early ages), the limb is more or less atrophied—a condition which is especially observable when the atrophy is confined to one side. I am not aware, however, that any microscopical observations on the nerve-centres exist supporting such a view in these cases.

If these atrophic conditions, either with or without talipes, be contrasted with the well-nourished condition of the limbs in the majority of cases of congenital talipes; and if the completeness of the recovery from the congenital deformity be further borne in mind the insufficiency of the nerve theory becomes, I think, abundantly manifest.

2. BONE CAUSES.—Many authors of repute have attributed club-foot to changes in the form of some of the tarsal bones, and in the direction of their articular surfaces. It would appear, however, that their observations were based almost entirely upon dissections of adult feet, in which secondary changes, due to walking and other causes, were mistaken for, or confounded with, such as may have been primary and present at birth. For this reason I shall pass over their writings, and come at once to modern authors on this subject. I will endeavour to trace whether the changes, which are undoubtedly present in many cases, are concurrent with, or secondary results of, the deformity.

Mr. WILLIAM ADAMS,* in 1852, described with great accuracy certain deviations in form and position of the astragalus met with in cases of congenital varus, and in his well-known Jacksonian Prize Essay,* he gave the results of investigations into the nature and bearing of these deformities, summing up his views thus :—" . . . the altered form of the astragalus, therefore, I regard as the result rather than the cause of the deformity."

* 'Club-foot; its Causes, Pathology, and Treatment,' 2nd edition, 1873, p. 159, *et seq.*

Hüter* subsequently corroborated Adams' description of these deviations, but instead of regarding them as a consequence of the malposition he considered them to be its initial cause. ". . . . congenital club-foot, in my opinion, consists chiefly in a pathological alteration in the form of the bones, and especially in the form and direction of the joint surfaces which (alterations), however, are closely allied to the physiological fœtal form." He believed that the articular facets of the astragalus in newborn infants tend naturally to place the foot in a supine position, while in the adult these surfaces have become altered in direction and tend to place the foot in a prone position. As regards the shortened muscles, he thought that this was a natural defect in their development, the length of a muscle depending on the distance between its points of attachment.

My own observations, both clinical and anatomical, coincide in the main with the descriptions given by these authors, and while I agree with Hüter that the deviations in form are of a physiological type, I think with Adams that they are a consequence of the malposition, and not its cause, as Hüter taught. Contrary, however, to the statement of Mr. Adams (op. cit., p. 153) that the neck of the astragalus at birth is normally continued directly forwards, I find, with Hüter, that the axis of the neck of the astragalus at birth is invariably directed forwards with a considerable obliquity inwards.

Although from the anatomical evidence it is clear that a capacity for inversion may be considered as physiological during fœtal life, nevertheless, the theory of malformation of the bones as the determining cause is

* 'Archiv für klinische Chirurgie,' vol. iv, part 1, p. 125, et seq.; and 'Virchow's Archiv,' vol. xxv, p. 598.

untenable for the following reasons: In all the cases of equino-varus (No. 2 excepted) there has existed a redundant portion of articular surface on the head of the astragalus to the outer side of the navicular bone, corresponding to the position occupied by the navicular in the normal condition as already described (p. 11). A similar redundancy is observable also on the anterior portion of the trochlear surface of the astragalus, corresponding to the position which would be occupied by the tibia if the foot were not in an abnormal position of extension. (In particular cases similar conditions are met with in other joint surfaces, e. g. in the knee-joint in " Genu recurvatum.") The facts mentioned undoubtedly show that displacement of the foot has occurred at a period subsequent to the development of an astragalus of normal conformation. In Case 2 the conformation of the astragalus was quite normal—a fact which argues strongly not only against malformation of the bone as the cause, but even as an essential element, of the condition.

Such variations in the conformation of the bones appear to indicate a difference in the period of onset in the two cases; it may reasonably be presumed that the deformity in Case 2 is of comparatively recent date, and has depended upon some accidental position of the limb, such as will presently be referred to. Moreover, in the anthropomorpha, this peculiar conformation of the neck of the astragalus exists throughout life as a normal condition, and yet these animals are not talipedic.

It would be out of place to describe in detail the various malpositions of the limbs, which may be present in the foetus or child at birth; it is nevertheless highly important to notice them, as they are strictly parallel with clubbing or malposition of the foot. The best and least uncommon example that can be brought forward is the knee-joint. It at times happens that the fœtus at term is packed with the hips fully flexed, and the knees extended, so that the feet lie opposite the head or face or neck. Mr. ADAMS (op. cit., p. 350), quoting LONSDALE, mentions such cases associated with breech presentation and talipes calcaneus.

That the position is not a passing one, or one assumed by the fœtus at the time of parturition in adaptation to the requirements of the process, is plain from specimens showing this position some time before the full term. In Guy's Hospital Museum there are two fœtuses, one at the seventh month within its membranes, exhibiting this position. A third specimen may be seen in the Museum of the Royal College of Surgeons, No. 3646A.

Such positions may be quite transient, being overcome after birth, partly by muscular action, and partly by manipulation. But, corresponding to that clubbing of the feet which is irremediable by nature, cases occur in which the extended position of the knees is rendered permanent by shortness of the muscles and ligaments and alterations in the joint surfaces. Mr. SHATTOCK has carefully dissected two such cases :—In the first case the hips admitted of only limited extension; the knees were over-extended and incapable of being brought beyond the straight line. After all the muscles had been dissected away, no flexion of the knees was possible, and it was only after dividing the capsule, in front of the lateral ligaments, that flexion became possible. The articular surface of the femur had undergone alterations in shape, in adaptation to the position of over-extension in which the joint was fixed; around the impression for the inner condylar surface of the tibia the cartilage of the femur was covered by a layer of lax connective tissue, like synovial membrane, and similar to that already described on the unused part of other articulating areas. In the second case the articular surface of the femur had undergone modifications in form, and the ligaments were so adaptively shortened as effectually to prevent flexion, even after the other soft parts had been dissected away. In both cases the limbs were symmetrically deformed. VOLKMANN has named this condition "Genu recurvatum."* Fig. 12 shows a typical case

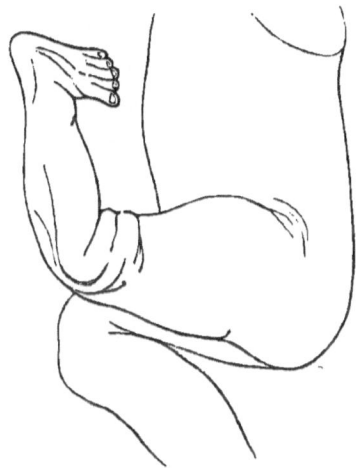

FIG. 12.—" Genu recurvatum." (Volkmann.)·

of the kind recently under my care in the East London Hospital for Children. A flexible splint was applied; and flexion of the knee-joints was gradually obtained. A similar condition of the knee-joints existed in the case figured on page 33 (Fig. 14).

The same is true of club-foot; it exists in all degrees and in all varieties. In the mildest cases the bones may

* For an interesting collection of these cases, see 'Congenital Malformation of the Knee-joint,' by Dr. J. F. HARTIGAN, Washington 1880.

present no alterations in form, and the natural use of the part may suffice to restore its proper position. Some limitation of extension-movement at the ankle-joint is quite common at birth; the foot does not admit of being flexed beyond a right angle. This condition is but a mild degree of talipes calcaneus, and is due to exaggeration of the flexed position of the ankle-joint which is normal during the later months of fœtal life. Such a physiological condition affects both feet, and is part of the same group of conditions as the inability to fully extend the knees or the hips, first demonstrated as constant in the newborn child by HÜTER, and all alike indicating a physiological shortness of muscles and ligaments in adaptation to the confined position of the limbs during intra-uterine growth.

The lesser severity of talipes calcaneus may be associated in part with the later period of its production, and in part with the anatomical arrangement of the affected joint. As contrasting with this, the condition of talipes equino-varus gives evidence of its early production in the great alterations in form of the astragalus and of the calcaneum; these are especially observable in the subdivision of the navicular facet of the astragalus, and the incurvation of the anterior part of the calcaneum; further evidence of early production is afforded by the shortness of the ligaments, and the adhesions sometimes found within the ankle-joint, as described in Cases 3 and 6, and of which another extreme case may be seen in St. Bartholomew's Hospital Museum (No. 3511). The lesser degrees of talipes equino-varus would seem to suggest that the original error in the environment has been transitory. The possibility of this is suggested by the different sources of the amniotic secretion; during the earlier period of intra-uterine life it is a transudation from the tisssues of the fœtus and a plexus of vessels formed beneath the amnion; during the later period these sources become obsolete, and its further increase is due to the urinary secretion of the fœtus. It is thus possible that an early deficiency in amount may be subsequently corrected, and that any mechanical effects, which may have been reproduced on the fœtus, will have time to right themselves, or if persistent will be but slight in amount.

3. MECHANICAL CAUSES.—In advancing the following views on the causation of club-foot, as already stated I do not deny the agency of other cases in a few exceptional cases. But for the great majority of cases I shall endeavour to show that club-foot at birth is a deformity only because it is a permanent, though sometimes exaggerated, condition of a physiological position.

We must go back to the writings of HIPPOCRATES for the earliest record of a mechanical causation for talipes; it is by him stated that children are deformed in the womb in consequence either of contusions, falls, or other accidents. Also, when the womb is relatively too small for the fœtus. It would serve no useful purpose to trace onwards the growth and development of this view: in recent times, CRUVEILHIER has been the most distinguished supporter and advocate of the mechanical theory. He taught that the deformity depended not so much upon direct pressure of the uterus on the fœtus as upon the pressure of one part of the fœtus on another. In his 'Anatomie Pathologique,' he figures a fœtus, the knees of which are extended, the right foot being locked beneath the chin in a position of extreme equino-varus, while the left foot, also in a varus position, rests upon the prominence of the left shoulder. FERDINAND MARTIN,* another warm supporter of the mechanical theory, was led to adopt it on being shown by DUPUYTREN a newborn infant "of considerable size and affected with a double inverted foot." On examining this infant attentively, he says, "we were struck by the position it at once assumed when left to itself—absolutely that which it had occupied in the womb. The thighs became flexed on the abdomen, the legs on the thighs, the feet applied themselves to the buttocks, and the whole body rolled itself up into an ovoid form." On making inquiries of the mother, MARTIN found that her abdomen had not been much distended during her pregnancy, and that the fœtal movements had been very painful; moreover, parturition had been very slow, and the "amount of the waters had hardly exceeded a couple of spoonsful." These facts gave MARTIN the notion that talipes depended on a constrained position of the *fœtus in utero*, and further, that this was the result of deficiency in the liquor amnii.

I can add a case from my own practice in every way parallel with the one just described. Fig. 13 is from a drawing made at the time the child was first seen at the hospital, and represents this intra-uterine position, which leads to double talipes equino-varus. The child for many weeks continued to assume this developmental position whenever it was left uncontrolled (as when naked) to do so. The talipes was not very severe and quickly yielded to appropriate treatment.

An abnormal intra-uterine position, also taken from life, is represented in Fig. 14; in consequence of undue pressure in the uterine environment a double and severe calcaneo-valgus has resulted. Besides the talipes there was fixation of the hip- and knee-joints, in the posi-

* 'Mémoire sur l'Étiologie du Pied-bot.' Paris, 1839.

tions shown in the drawing, a condition which I have endeavoured (p. 29) to show is the analogue of club-foot, due to the same cause, and having analogous anatomical features,—shortness of the ligaments. This child at first

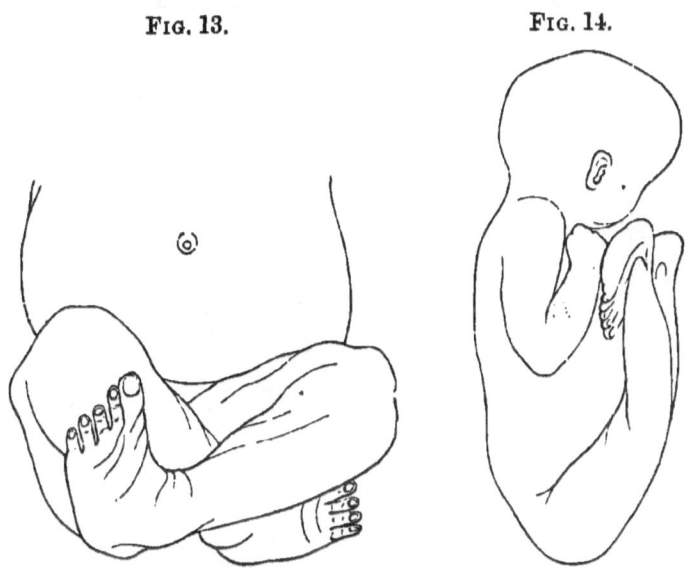

FIG. 13.

FIG. 14.

FIG. 13.—Typical intra-uterine position, leading to equino-varus.
FIG. 14.—Abnormal intra-uterine position, leading to calcaneus.

could assume no other position, and for many weeks after its birth, when naked, the limbs reverted more or less to the position shown in the drawing.

In the course of his practice, MARTIN met with not less than sixty cases of club-foot due, as he thought, to deficiency of the amniotic fluid, and finally he arrived at the conclusion that though probably not the sole, it was the most frequent cause of the deformity. In this paper, which is one of great interest, he relates and figures a variety of cases. I have selected the two following figures as showing normal positions favouring the development of one-sided equino-varus, and one-sided calcaneus. The fœtus being in one or other of these positions, it will be obvious how any slight pressure from the enveloping uterus will fix the feet, even tend to exaggerate the position, and so lead to the production of talipes.

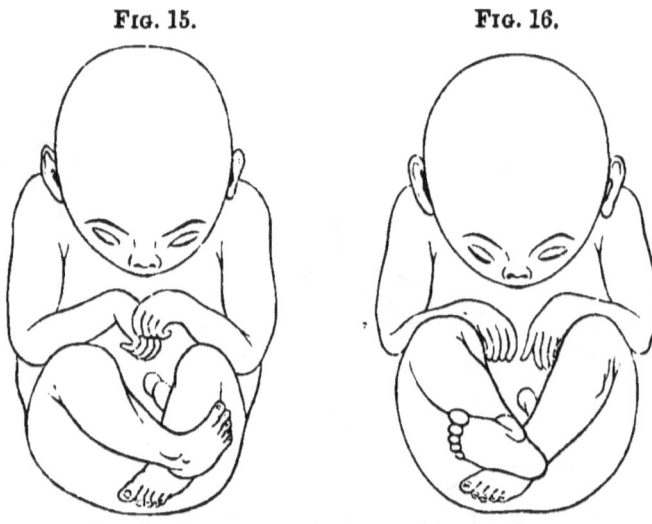

Fig. 15.—Normal developmental position, favouring production of equino-varus.

Fig. 16.—Normal developmental position, favouring production of calcaneus.

The two commonest forms of the deformity, extension of the ankle with inversion of the foot (T. equino-varus), and flexion of the ankle with eversion of the foot (T. calcaneo-valgus), represent the two commonest physiological movements of the foot. The former position is natural to the fœtal feet during the early period of intra-uterine life, the latter during the later period. Equino-varus is more common and more fundamental as a deformity than calcaneo-valgus, because the causes which produce it begin to act earlier, and are therefore longer maintained.

MY ARGUMENT is that the feet should occupy various positions during the course of intra-uterine life; this is doubtless ordained in order that the muscles, the joint-surfaces and the ligaments shall be so developed as to allow of that variety of movement and of position which are afterwards to be natural to the foot. Should anything prevent the feet from assuming these positions at the proper time, or maintain them in any given position beyond the limit of time during which they should normally occupy such position, a talipes results. The severity of the deformity will be in direct ratio to the violence at work; the variety of the deformity will depend on the period when this violence first commences to act.

There is abundant pathological material to show that talipes may be brought about by mechanical means in several apparently distinct ways. Without pretending to have exhausted the subject, the following appear to be the more common modes of production :

(1) By accidental locking of parts.
(2) By locking of parts due to abnormal position of the limbs.
(3) By exceptional positions of the limbs independently of locking.
(4) By uterine environment from actual or relative deficiency of liquor amnii.
(5) By congenital absence of certain bones.

These causes will now be considered *seriatim*. It may be mentioned that talipes has been said to have resulted from tight lacing and from modes of dressing intended as long as possible to hide the pregnant condition. I can only say that I have no experience of such cases, and that I know of no pathological specimens bearing on these points.

The following case occurred in the practice of Dr. Biggs, of Wandsworth, to whom, through my friend Mr. Shattock, I am indebted for the notes, and for permission to publish them.

> A married lady came under observation in February, 1885, for what appeared to be a tumour of the uterus occupying Douglas' pouch; the tumour was "reduced" and prevented from returning to its position by a pessary; it was at this time as large as the uterus at the fourth month of pregnancy. About July of the same year the lady became pregnant. The tumour increased in size and came to occupy the left side. Pregnancy was attended with great distress, owing to the presence of the tumour; it terminated in February, 1886. Labour was natural; there was not much liquor amnii; forceps were used to expedite delivery and on account of the presence of the tumour. The child at birth presented a well-marked talipes calcaneus of the right foot; there was a depression on the left side of the neck, and the head was held inclined to the right side for about a fortnight after birth.
>
> The talipes disappeared about seven weeks after birth, without any treatment beyond manipulation. It may be concluded, therefore, that the right foot was applied to the left side of the neck, and pushed the head over to the right. Sir Spencer Wells, who saw the case, made the prognosis that the deformity would entirely and spontaneously disappear, which it did in the course of four or five months.

(1) By Accidental Locking of Parts.

It will be remembered that the limbs arise from the lateral parts of the trunk, as semi-lunar plates of the parietal mesoblast with its investing epiblast; and present, each, a dorsal (extensor) and a ventral (flexor) surface, the thumb and great toe being towards the head or pre-axial. As development takes place, the primary lappets, from being at first simple lateral extensions of the trunk, come to be folded ventrally or against the body of the embryo; the anterior with something of a backward, the posterior with something of a forward direction. The early subsequent alteration in position of the hind limb necessitates a rotation from the hip-joint downwards, by which the extensor surface of the thigh, leg, and foot are brought forwards, and their flexor surfaces carried backwards. The late Professor ESCHRICHT,[*] of Copenhagen, was the first to study these developmental movements and positions in relation to deformities of the legs and feet. He would not, however, allow that interference with these movements—on which he recognised the dependence of deformity—was due to intra-uterine pressure, but attributed it rather to "arrest of development." Many subsequent writers (BILLROTH, VOLKMANN, LÜCKE) have acknowledged the value of ESCHRICHT's observations, while Dr. BERG,[†] of New York, in a paper on the "Etiology of Congenital Talipes Equino-varus," appears to have independently arrived at a similar conclusion, viz. that club-foot depends on a retarded or on a non-rotation inwards of the lower extremity; but, like ESCHRICHT, he has not attempted to explain on what this non-rotation itself depends. In the 'Lancet' of January 20th, 1887, I have replied critically and at some length to Dr. BERG's views.

An examination of well-preserved fœtuses, with the membranes still intact, shows that flexion of the knee and inward rotation at the hip-joints occur at very early periods quite independently of the environment, as may be inferred from the large proportional amount of liquor amnii. By the increase in length of the lower limbs the feet, which at first were apart, come to meet over the lower part of the abdomen, and subsequently to cross, the normal position at this period, as may be well seen at

---

[*] 'Deutsche Klinik,' No. 44, Nov., 1851, pp. 467—470, "Ueber die Foetalkrümmungen, namentlich in Bezug auf die Bedeutung der angeborenen Verdrehungen der Bauchglieder."
[†] 'Séguin's Archives of Medicine,' vol. viii, No. 3, Dec., 1882.

about the fifth or sixth week, being one in which the hip- and knee-joints are flexed and rotated outwards, and the crossed feet inverted, of which specimens may be seen in St. Bartholomew's Hospital Museum, No. 1205; and in St. Thomas's Hospital Museum, No. 1348.

From other specimens, however, it would appear as if this crossing of the limbs were occasionally interfered with. Thus, in a specimen (St. Mary's Hospital Museum, M. $\frac{c}{\text{ii}}$, North Collection), that of a well-nourished fœtus within its membranes, of about eleven weeks, the soles appear to have met so exactly as

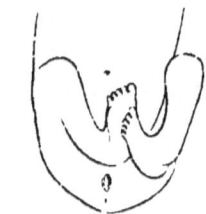

FIG. 17.—Accidental locking of feet, in a normal position of limbs.

to interfere with the crossing of the legs; the feet, pressed against each other, are slightly flexed (talipes calcaneus), the sole of the right foot is arched, and receives the anterior part of the left, and its outer border is everted. There is abundant space within the membranes. This is the only unequivocal specimen I have been able to find illustrating club-foot at so early a date in healthy fœtuses within membranes distended with the normal amount of amniotic fluid; and the deformity appears almost certainly due to the accidental locking of one part against another, independently of surrounding pressure. Regarding this specimen, however, it is important to note that its age is computed at eleven weeks, the period at or about which, according to modern investigations,* the fœtus first begins to employ its muscles; and it is hardly possible to believe that such a position could be long maintained after the period at which movements commence.

(2) BY LOCKING OF PARTS, DUE TO ABNORMAL POSITION OF THE LIMBS.

Among the best examples of the locking of one part in another resulting in marked deformity may be mentioned the case described and figured by CRUVEILHIER; both knees were extended; the right foot was locked beneath the chin, and had, during the growth of the limb, gradually assumed a position of extreme equino-varus. It is worthy of remark that both hands were clubbed, and that the condition was associated with absence of the radius from each forearm.

Cases of this kind are also described and figured by MARTIN; one of the most interesting of these is represented

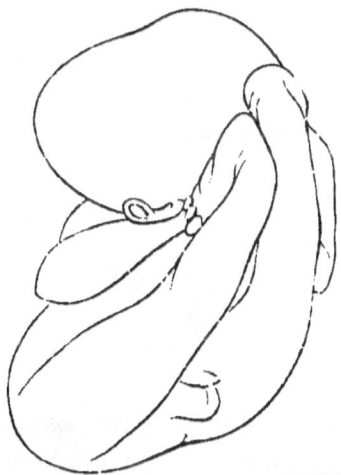

FIG. 18.—Locking of feet, in abnormal position of limbs.

* LUSK, 'Science and Practice of Midwifery,' p 99.

in Fig. 18. It was that of an infant (op. cit., p. 30), whose legs had been fully extended; the feet were flexed, and the right one was forcibly everted; the sole rested on the corresponding cheek, which was indented by it; the heel was pressing into the orbit, the globe being atrophied in consequence. The left limb was in a nearly similar position. Cases of the kind are also recorded by VOLKMANN (op. cit., p. 691); this author figures a case of talipes calcaneo-valgus on one side caused by pressure of its fellow foot, which is in a position of well-marked equino-varus.

It is interesting to note that MARTIN's case, which I have just quoted, was one of extra-uterine fœtation. Though the uterus cannot therefore be invoked as the

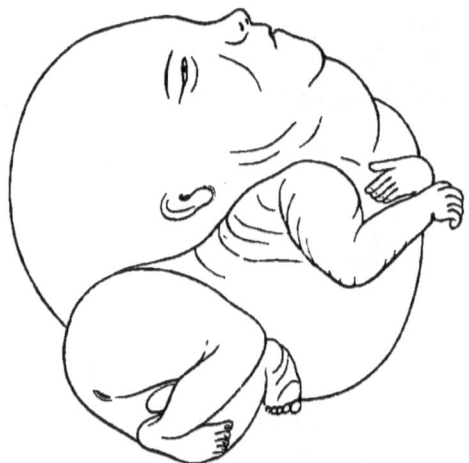

FIG. 19.—Effects of environment, in extra-uterine fœtation.

cause of the deformity, it is none the less clear that environment has brought it about. For another interesting case of deformity of the feet associated with extra-uterine pregnancy, I am indebted to my friend Dr. JOHN WILLIAMS. Fig. 19 represents the fœtus, removed by Cæsarian section in the thirty-fifth week of pregnancy; the mother made an excellent recovery. In this case, the deformity differed on the two sides; the right sole and heel looked directly inwards, the ankle-joint being fully flexed; the left sole looked directly outwards, the ankle-joint being fully flexed, as on the right side. Besides the position of the feet, there were many other evidences of the mechanical action of environment on this fœtus.

(3) BY EXCEPTIONAL POSITIONS OF THE LIMBS, INDEPENDENTLY OF LOCKING.

That exceptional positions of the entire limb *in utero* may lead to the production of talipes, independently of

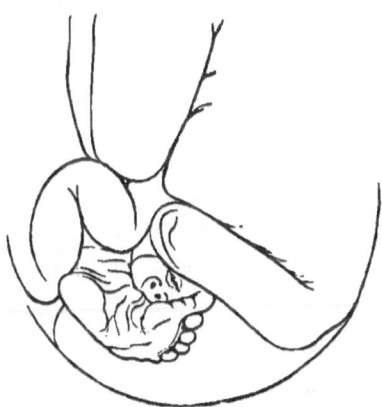

FIG. 20.—Fœtus with equino-varus, from malposition of limb.

locking, is well shown by a specimen in St. Bartholomew's Hospital Museum, No. 1226, and is represented in Fig. 20.

It is a healthy, well-nourished fœtus of about seven months, in position within the membranes and uterus; the right lower limb is in a normal position; on the left side the thigh is flexed, but the knee is extended, and the foot lies on the opposite cheek, in a position of marked varus, with the sole and heel opposed to the wall of the uterus. In this case there can hardly be any doubt that the talipes is the result of an unusual position of the limb for which uterine environment can hardly be held responsible, although when once established, environment would doubtless tend to maintain. This case differs considerably from the preceding one, in that there is no locking of parts.

(4) By Uterine Environment from Actual or Relative Deficiency of the Liquor Amnii.

Martin came to the conclusion (op. cit.) that a deficiency in the waters was the commonest of all causes. He observed not less than sixty cases in his own practice caused, as he thought, in this manner. I cannot, however, bring myself to accept this doctrine in its entirety, for there are many well-authenticated cases in which no such deficiency, but the contrary, has been observed. I have convinced myself, nevertheless, by careful inquiry that the fœtal movements have been unusually felt, and have even caused much pain to the mothers in quite a proportion of cases of club-foot, and that the amount of the waters has been smaller than natural. I should have little doubt that this pain experienced during pregnancy, as well as the deformity present at birth, were due to the limitation of uterine space from deficient liquor amnii.

The known existence of club-foot at early periods of fœtal life has been used as an argument against the mechanical theory of its causation, it being assumed that uterine pressure could have no influence in presence of the relatively large amount of liquor amnii present at such periods. On the other hand, the occurrence of adhesions between the amnion and the surface of the fœtus are incontrovertible evidences of the occasional close apposition of the amnion and the fœtus, and the very early stage at which this occurs. There are many interesting specimens in the London museums, one of the best being No. 3059, St. Bartholomew's Hospital Museum. In lesser degrees amniotic bands are of frequent occurrence, and are not improbably the cause of many intra-uterine amputations, withered limbs, and other malformations met with in infants.*

The ordinary intra-uterine position in double equino-varus is represented in Fig. 21 from a preparation in

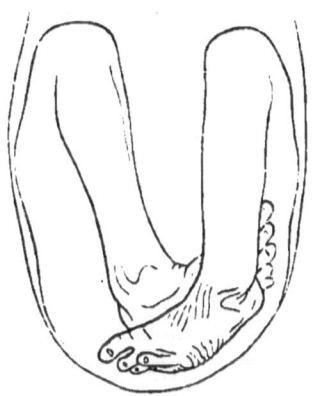

Fig. 21.—The intra-uterine (developmental) position in double equino-varus.

St. Mary's Museum (M $\frac{A}{2}$ North Collection), though the disposition of the lower limbs in cases of varus is one on which further data are highly desirable. In this specimen, a well-preserved fœtus within its membranes of four or five

* For an interesting account of this subject, including a list of Continental specimens and published cases, see the Inaugural Thesis of Dr. KARL KLOTZ, ' Ueber amniotische Bänder und Fäden,' Leipzig, 1869.

months, the position of the hips and knees is normal; the sole of the right foot rests on the opposite buttock; the left foot crosses the dorsum of the right; both feet are in a position of well-marked equino-varus; the fourth and fifth toes of the left foot are strongly adducted over the roots of the second and third; indeed, both feet exhibit signs of considerable compression. In keeping with this pathological fact the broad flattened-out appearance of many club-feet will be borne in mind.

A yet more palpable effect of intra-uterine pressure is rotation of the long bones. I recently dissected a specimen, which had been removed from a fœtus at term, with a double talipes equino-varus. A short distance above the left ankle-joint there was a deep linear depression encircling the limb, as though it had been constricted by a firm amniotic band. On removing the integument the deeper structures were found unimplicated; the whole limb was rotated on its long axis from within outwards; the fibula was completely behind the tibia; the internal surface of the tibia, including the inner malleolus, presented directly forwards, and was, in fact, covered by the inner border of the gastrocnemius muscle; the astragalus was partially extended; the os calcis was rotated inwards, and its greater tuberosity looked inwards and somewhat upwards. The internal lateral ligament was short, strong, and held the internal malleolus in close apposition with the astragalo-scaphoid articulation.

Besides the DIEFFENBACH'ian form—the physiological club-foot with which so many children are born—the curving of the leg bones may be mentioned as another instance of developmental posture. It not infrequently happens that so much curving of the tibiæ is present at birth as to prevent the internal condyles of the femurs from touching when the feet are brought together. I have several times been consulted on the desirability of applying splints in such cases.

MARTIN states that twin children are frequently born with some deformity or other, especially club-foot, and accounts for it on the theory of mutual interference with each other's position, and of a relatively small amount of the liquor amnii, which he asserts is not in excess of that present in the case of a single child. I have under treatment at the present time a twin child suffering from congenital equino-varus of the left leg; the other child is normally formed in all respects. I have had one or two other cases in the past, but in the great majority of talipedic children no such cause could be invoked.

Good evidence of the effects of compression is further afforded by the hand represented in the adjoining figure, for which I am indebted to Dr. SILCOCK. Both hands were pressed flat, as shown in the woodcut,

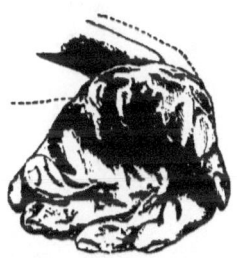

FIG. 22.—Fœtal hand, showing marked signs of pressure.

between the thighs and the abdomen, which was greatly distended by cystic kidneys. There was double talipes equino-varus, and the hip-joints were in a condition of subluxation.

The presence of areas of atrophied skin with bursæ beneath, situated over the most prominent points of the exposed aspects of the feet further confirms the view, that the deformity is a condition enforced by the mechanical action of the environment. These atrophic areas were first described by VOLKMANN* and subsequently by LÜCKE,† two distinguished advocates of the mechanical theory. In Dr. WILLIAMS' specimen just mentioned, there was a shiny area of atrophic skin over the right external malleolus.

(5) BY CONGENITAL ABSENCE OF CERTAIN BONES.

Deformities resulting from the absence of bones are so exceptional that little need be said about them. They follow no very definite rule as far as I can make out. The drawing (Fig. 23) represents a symmetrical displace-

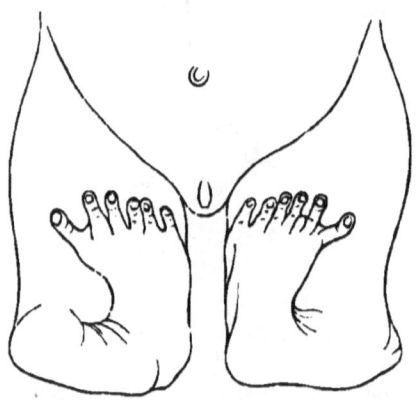

FIG. 23.—Congenital absence of the tibiæ.

ment of the feet inwards—talipes equino-varus; it was taken from a child two years of age, whom I was enabled to show at the Pathological Society‡ a year or two ago,

* PITHA and BILLROTH, 'Handbuch der Chirurgie,' vol. ii, p. 690.
† "Ueber den angeborenen Klumpfuss," 'Klinische Vorträge, No. 16.
‡ 'Transactions,' 1882, vol. xxxiii, pp. 236-8.

through the kindness of my friend, Mr. HENRY BAKER. A similar deformity dependent on partial absence of the tibia is shown in Fig. 24.

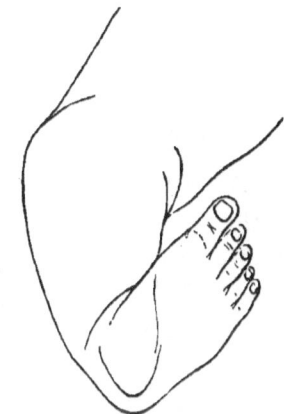

FIG. 24.—Congenital absence of the lower portion of the tibia.

In November, 1880, Mr. Gould showed at the Pathological Society a boy three years of age, suffering from a congenital talipes equino-valgus on the left side due to absence of the fibula. In this case section of the tendo Achillis proved highly serviceable; after being provided with a boot and iron support on the outer side the boy was able to walk with ease.

The comparative rarity of club-hand apart from deficiency of bones, as explained by CRUVEILHIER, is probably due to the anatomical differences in the carpal and tarsal articulations, the great amount of mobility of which the hand is capable serving to obviate the serious effects which would be due to malposition in a member having a less range of movement. It may be suggested further that the position of the hands in the space between the thighs and the head serves as a protection against immediate pressure.

A well-marked instance of congenital club-hand not dependent on deficiency of bones is depicted in Fig. 25. On the right side the wrist was almost fixed in the flexed position; the forearm slightly pronated; the thumb was fixed in adduction; the fingers were slightly flexed; the hand was small but broad in proportion; the left hand was larger and broader than the right; the thumb was in contact with the palm and could not be abducted; the forearm was more pronated, but the wrist less flexed than on the right side. There was double equino-varus, which quickly yielded to treatment.

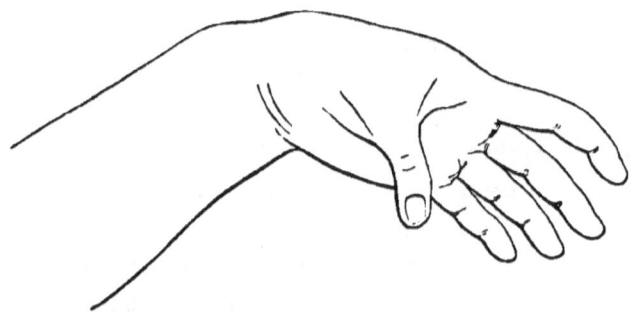

Fig. 25.—Congenital club-hand, from rigidity of ligaments and muscles.

The more common form of club-hand depends on absence of the radius, partial or complete, and on deficiency in the carpal bones. These cases do not follow any definite type, and vary much among themselves. Such a condition is represented in Fig. 26 from a case shown at the Clinical Society, and recorded in the 'Transactions' for 1886. Though the two arms are very similar in appearance the anatomical structure varies. In one there is complete absence of the radius, in the other the radius can be felt, though in a very dwarfed condition. The humerus may or may not take part in the deformity; the carpus may be complete or its radial portion more or less deficient; the thumb may be absent, rudimentary, or pedunculated; the first and second fingers, one or both,

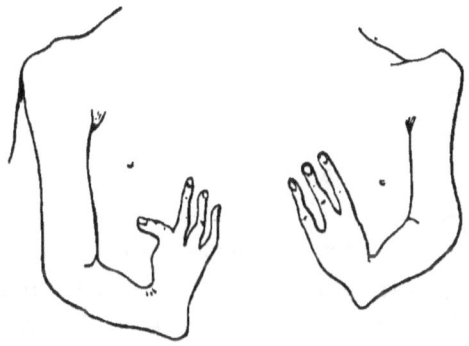

Fig. 26.—Congenital club-hand from absence of bones.

may be absent or present. This want of uniformity suggests that club-hand is a condition merely associated with ill-development of the forearm, while there are morphological reasons for not regarding it as a reversion. In the light of DARESTE's* experimental inquiries into the artificial production of monstrosities I may claim these cases, not without good grounds, as additional evidence of the effects of environment on the fœtal skeleton.

LITTLE (op. cit., p. 24) argues that: "the occurrence of the perfectly analogous deformity of the hands which takes place prior to birth, denominated club-hand, in which the flexors and pronators (analogous to the so-called extensors and adductors of the foot) are likewise contracted, corroborates the opinion that congenital club-foot

* 'Recherches sur la Production Artificielle des Monstruosités, Paris, 1877.

depends on spasmodic muscular contraction." I venture to take an exactly opposite view. If the deformity depended on either a paralytic or a spasmodic condition of the muscles, how is it that club-foot is so very much more common than club-hand? It would probably be within the truth to say that one thousand cases of the former occur to a single one of the latter.

The foregoing classification of causes is somewhat absolute, but it will serve to emphasise what I believe to be the predisposing cause in each class of cases. In any given case it is difficult to say how far the original and primary deviation from the normal development or position is accidental, and how far it depends upon limited space within the uterus. In the earlier weeks there can be little doubt that the position of the fœtus is assumed in response to an inherent power of development; somewhat later on, muscular movements on the part of the fœtus, assisted by uterine contractions, are the recognised means by which position may be determined. These movements, however, must be very slight, and very slight causes will suffice to interfere with them. As the fœtus develops the influence of environment becomes more possible in consequence of a diminution in the uterine space relatively to the size of the fœtus. In proportion as this pressure is severe, continuous, or intermitting, will be the degree of severity of the resulting lesion. As regards the feet it may suffice to say that deformities take place "according to certain invariable rules" (LITTLE), because of the anatomical structure of the foot. The commonest deformity of all, that of equino-varus, is the position of perfect rest; it is the position into which the foot naturally falls when uncontrolled by muscular action, as during deep sleep, or chloroform narcosis, and which needs the least expenditure of force for its maintenance.

It must not be supposed, when invoking environment, that some great mechanical power is referred to. If the extreme delicacy of the growing fœtus be remembered, and the nature of the majority of the deformities—a more or less exaggerated condition of a sometime physiological type—it will be conceded that but little pressure is needed. And that the uterus may and does exercise pressure on the fœtus within it, there are numberless examples, as already stated, other than club-feet, to show. Rather should we be surprised that nature errs so seldom, and that so few malformations occur, notwithstanding the accumulated weight of hereditary influence, the conditions of life of many thousands of child-bearing women, the direct injuries they receive, and the utter neglect of personal health during the most critical periods of their pregnancy.

With regard to heredity, the whole question is so obscure that it is hardly profitable to enter upon it. Although cases of hereditary transmission of the deformity from parents to children do undoubtedly occur, yet in the vast majority of instances no such hereditary influences can be traced; and this is even true in instances where several children of the same family have suffered. The influence of heredity, however, may be invoked with equal force, whatever view of the pathology of the disease be adopted. But at first sight it may seem not a little remarkable that in some cases the deformity is transmitted along the paternal line, and it may be difficult to harmonise this fact with the influence of environment on which we have insisted. It need only be remarked, however, that the environment of the fœtus depends upon the fœtus itself, not less than upon the mother. For the most recent observations show that the liquor amnii may be considered throughout a fœtal and not a maternal product. Excess or deficiency in its amount may therefore be the result of a tendency inherited either from the father or the mother.

CHAPTER III.

VARIETIES OF CONGENITAL CLUB-FOOT.

Anatomical classification—Talipes calcaneus and calcaneo-valgus—Talipes equinus and equino-varus—Irregular forms—Anatomy of each variety:—Ligaments, muscles, bones, skin—Untreated and relapsed forms.

HAVING tried to show that congenital talipes is a structural fixation of the foot, with or without exaggeration, in one or other of its normal developmental positions, I will now describe the varieties of club-foot as met with at the time of birth or shortly afterwards; that is to say, before secondary changes from walking or other causes have had time to develop. Club-foot admits of a strictly anatomical classification, and the commonest form corresponds to that position which the fœtal foot earliest assumes and longest occupies.

The deformity corresponding to flexion of the ankle-joint (T. CALCANEUS) occurs rarely. For anatomical reasons flexion of the ankle is associated with some degree of eversion of the foot, and it is to be noted that the flexion deformity rarely occurs without eversion (T. CALCANEO-VALGUS). Eversion (T. VALGUS) also rarely occurs alone, for the reason that any force acting on the outer part of the foot would tend first to flex the ankle-joint, and then, if continued, to exaggerate the natural inclination of the foot during flexion to become everted. Thus we get

T. CALCANEO-VALGUS as the usual type of a flexion deformity.

The deformity corresponding to extension of the ankle-joint (T. EQUINUS) is also exceedingly rare as an uncomplicated congenital form, for the reason that, when extended, the foot is usually somewhat inverted. Most recent authors refer to inversion of the foot (T. VARUS) pure and simple. I have, however, never met with a case. Talipes varus, so called, even when there is no pointing of the toes, is always found associated with extension of the ankle-joint. Mr. SHATTOCK has made several dissections, and has invariably found this to be the case. So we get the commonest of all forms of congenital club-foot, T. EQUINO-VARUS. The inversion is chiefly brought about by rotation inwards of the os calcis on the astragalus, in some measure also probably by movement of the fore part of the foot beyond the transverse tarsal-joint.

Thus we meet with the following congenital forms :—

    (1) TALIPES CALCANEUS (flexion of the ankle-joint, very uncommon). T. CALCANEO-VALGUS (flexion of the ankle with eversion of foot).

    (2) TALIPES EQUINUS (extension of the ankle-joint, very uncommon). T. EQUINO-VARUS (extension of the ankle with inversion of the foot).

This last form constitutes at least 90 per cent. of all the congenital cases requiring surgical treatment. Besides these chief forms, other deformities are occasionally met with which cannot so strictly be brought within the anatomical definition. They appear to me to strengthen the view of mechanical causation, being irregular positions such as muscular action could not bring about. The extension deformities are much severer than any of the others, for the reason, I believe, that they commence earlier in intra-uterine life, at a time when the fundamental form of every element in the foot is most capable of being influenced by any, even very slight, unusual environment acting upon it.

TALIPES CALCANEUS.—In its milder forms calcaneus with slight eversion of the foot is quite a common condition. If the feet of new-born children be examined it will be found, in a certain number of cases, that the ankle-joints can be flexed, so that the dorsum pedis will lie upon, or almost upon, the leg, and that it cannot be extended beyond a right angle. This condition, normal at birth,

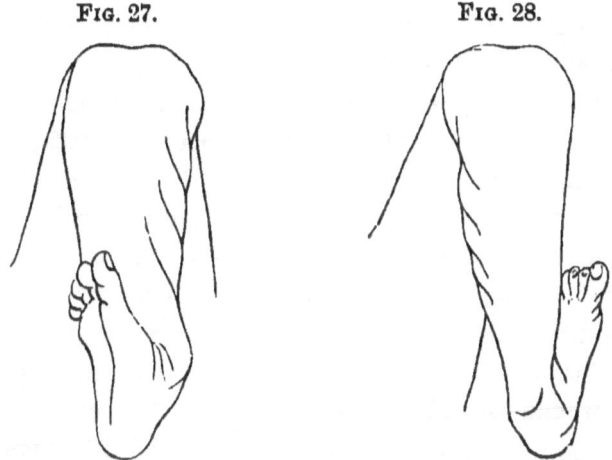

FIG. 27.   FIG. 28.

FIGS. 27 and 28.—Congenital calcaneus. It will be observed that the feet are everted as well as flexed. From an infant four weeks old.

gradually changes, so that flexion becomes more and more limited, and extension more and more increased. In one or two cases out of every hundred, the degree of calcaneus amounts to a deformity owing to shortness of the anterior ligament of the ankle-joint and of the anterior muscles of the leg.

Such a case is well illustrated in the adjoining woodcuts (Figs. 27 and 28).* The patient was a girl, and, at the time I first saw her, aged four weeks. She was a twin-child, the other being perfectly well formed. There was but little movement at the ankle-joint, not more than some 12° or 15°. It will be observed that the feet are everted; as has already been said, eversion to a greater or less degree is the rule.

The anatomy of talipes calcaneus was referred to in describing the cases in Chapter I. In Case 4, p. 13, the ankle-joint could not be extended beyond about 45° even after division of the tibialis anticus, and extensor proprius hallucis muscles; it was necessary likewise to divide the anterior ligament of the ankle-joint in order to secure complete extension. An almost similar condition was found in Case 5, p. 14. In both cases the calf-muscles appeared too long after extension had been obtained; but in other respects their structure was quite normal.

CALCANEO-VALGUS.—For the reasons given, calcaneus is most frequently associated with eversion of the foot. Beyond this, however, there is sometimes eversion of the outer border of the forepart of the foot, with a compensatory prominence of the inner border, the centre corresponding to the under aspect of the astragalo-scaphoid articulation. Such a condition is well represented in Fig. 29; also on page 34 in Fig. 16. The former drawing was taken from Caroline N—, aged 4 months. The ankle could be flexed to something short of a right angle without much difficulty, but the foot immediately relapsed

* This and several other of the drawings were made for me from children attending the East London Hospital for Children by Miss REEVE, one of the nurses at that institution. I take this opportunity of thanking her for them, and for the great readiness she has always displayed to help me in these cases. Her drawings have been reduced to their present dimensions by Mr. SHATTOCK.

when left to itself. The anterior fasciculus of the external lateral ligament, and the outer part of the anterior ligament of the ankle-joint, both became tense when extension was attempted. The dorsum of the foot over the

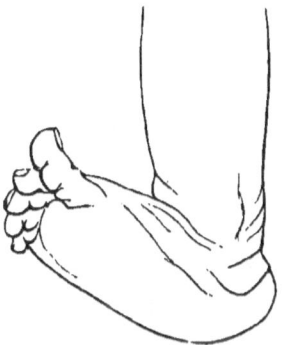

Fig. 29.—Congenital calcaneo-valgus.

cuboid was indented by contact with the outer malleolus and front of the leg. In a dissected case the tendo Achillis became quite flaccid when the foot was extended.

TALIPES VALGUS is recorded by some authors as a congenital deformity. I can only say that I have never seen an uncomplicated case in a young infant. Spurious forms, occurring quite early in life in rickety children, are common enough, but of these I need not further treat in this place.

TALIPES EQUINUS.—This is so rare a form of congenital deformity that it may be dismissed with a mere mention. I have only seen one case that could with any propriety be called equinus; this is represented in Fig. 30. It will be observed that the foot is somewhat inverted, as well as extended. The inversion was easily overcome; and the whole deformity yielded quickly, after section of the tendo Achillis.

TALIPES EQUINO-VARUS.—This is the common deformity,

to which when not otherwise qualified, the term "club-foot" always applies. Some authors use the simple term VARUS for this variety of club-foot. I prefer the compound word EQUINO-VARUS as conveying a more correct idea of the anatomical condition than the simple term VARUS.

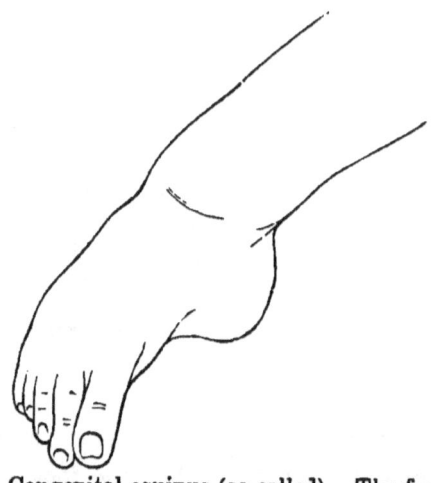

FIG. 30.—Congenital equinus (so called). The foot is also inverted.

Like the preceding conditions, it may exist in almost any degree of severity. As already stated, the normal foot at birth enjoys a far wider range of movement than at any other time afterwards; its capability for inversion is very considerable, and in not a few cases the soles habitually assume a position facing each other. Hence DIEFFENBACH's paradox that "all children are born with some degree of club-foot." This movement of inversion is due partly to the non-development of the *sustentaculum tali*, and partly to the pliability and elasticity of all the structures of the foot, including the cartilaginous basis of the tarsus. As the foot grows and as ossification proceeds, this movement of inversion becomes more and more limited.

In a proportion of cases, the capability for inversion passes beyond the normal and becomes a first stage of talipes. In such cases, the foot can be easily replaced, but when left to itself it will again revert back to its deformed position. The second stage may be described as one in which the foot cannot be even temporarily replaced in its normal position without the application of such force as will cause pain. The third stage of

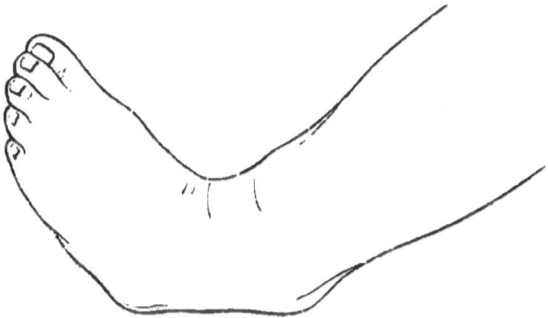

Fig. 31.—Congenital equino-varus. Second degree.

deformity may be said to be reached when replacement of the foot can only be accomplished after surgical division of the resisting structures. The second stage, if neglected, may run into the third, and the third may exist in varying degrees of severity even at birth. The annexed drawings are from infants, or young children before the age of walking, and fairly represent the uncomplicated forms of talipes equino-varus, of different degrees of severity (Figs. 32—35).

These primary deformities may undergo various changes, and any degree of secondary deformity may come on either from neglect of, or from walking upon, the deformed feet. One of the commonest and most striking changes noticed is atrophy of the leg, and, in a measure, of the entire limb.

VARIOUS FORMS OF EQUINO-VARUS.

Although patients hobble about and even walk long distances, the muscles of the limb do not develop; there is an entire absence of calf, and as one set of muscles is fixed by reason of being too short, so the antagonists are useless by reason of having none to antagonise.

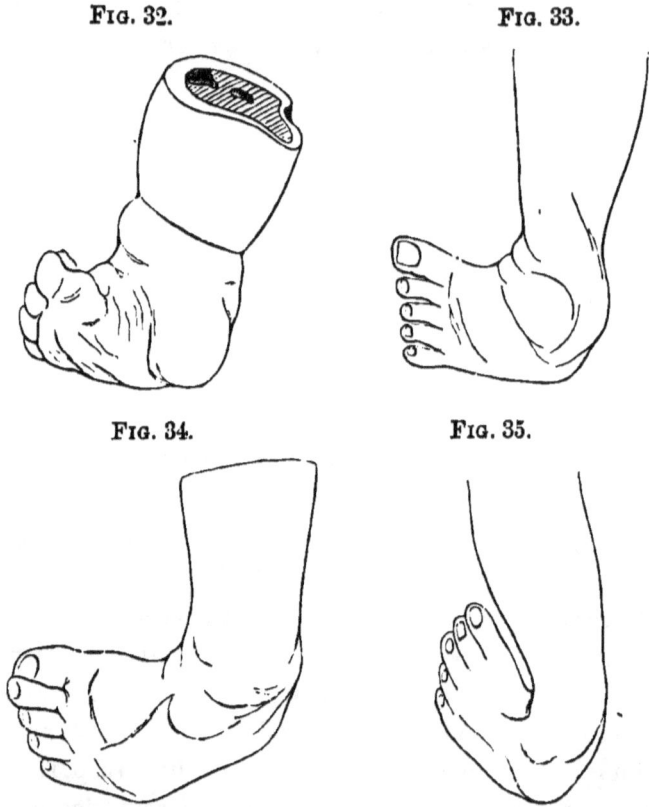

FIG. 32.   FIG. 33.

FIG. 34.   FIG. 35.

FIGS. 32—35.—Congenital equino-varus. Varying degrees of severity.

In severe cases of equino-varus it is not unusual to find the foot so much inverted that the sole looks upwards and inwards, while the dorsum is used to stand upon (Fig. 36). If neglected, such cases assume very intractable forms, which almost baffle treatment. Specimens of such feet are to

ANATOMY OF EQUINO-VARUS.

be found in every museum, and are figured in all works on orthopædic surgery. They are interesting as showing what amount of deformity the foot is capable of under-

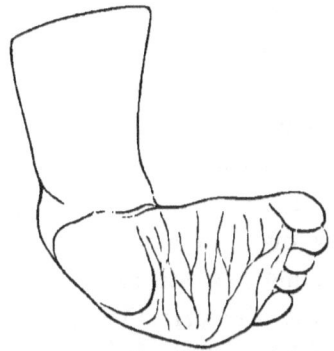

FIG. 36.—Severe equino-varus, from a child a few hours old.

going, but teach nothing as to its real anatomy. The cases differ among themselves according to their age and other attendant circumstances.

THE ANATOMY OF EQUINO-VARUS varies very much; the deformity is essentially a double one—extension of the ankle-joint with inversion of the foot, either of which may predominate. Inversion is more complicated than extension; it is a combination of rotation inwards of the calcaneum, the interosseous ligament acting as a centre, and of inversion of the component parts of the foot at or beyond the transverse tarsal joint. In many cases there is inversion of the leg as well as of the foot, and even of the entire limb, a condition for which CRUS VARUM is the most appropriate term. In the intra-uterine position which leads to equino-varus (Fig. 13, p. 33) it will be observed that the thigh is still partially rotated outwards, so that the patella, instead of presenting forwards, looks outwards. As the thigh gradually assumes its normal position after birth, the foot becomes more and more inverted; in severe cases the inversion may be such that the toes point directly backward. In estimating the degree of talipes in a very young infant, an attempt should always be made to place the patella directly forwards, or otherwise a misconception as to the severity of the talipes may result. In children who have learned to walk the rotation of the limbs will be completed and the degree of talipes more obvious. Cases of crus varum are very difficult to deal with, as there is an abiding tendency in the leg to turn inwards.

The anatomy of equino-varus has been described so often that I should hesitate to enter upon it here, were it not that some of the principal points have been either passed over in silence or so insufficiently dealt with that their bearing on treatment has been underrated. I refer now more especially to the condition of the tarsal ligaments. My attention has been prominently directed to this part of the subject during the past four or five years, in consequence of discovering, while dissecting some talipedic feet, (1st) that after all the muscles had been entirely removed the feet could not be put straight without exercising very much more force than could be surgically applied to, and tolerated by, the living foot; and (2nd) that on dividing certain of the tarsal ligaments, reduction of the deformity became quite easy, even when the muscles were allowed to remain *in situ*. All the component structures of the foot, however, are concerned in congenital club-foot. I will, therefore, consider these structures *seriatim* ; the ligaments, certain muscles of the leg, the bones of the the tarsus, and the skin.

THE LIGAMENTS.—Of first importance are those about the astragalo-scaphoid joint. In these cases, there is a capsule made up above and internally by a blending together of the superior astragalo-scaphoid ligament with fibres from the anterior ligament, and the anterior portion of the deltoid ligament of the ankle-joint; below, with fibres from the inferior calcaneo-scaphoid ligament. To these are united fibrous expansions of the tendons of the anterior and posterior tibial muscles; together they form an unyielding capsule of great strength, which is attached to the several bones, not in the usual manner, but in adaptation to their altered relative positions. This I would name the "astragalo-scaphoid capsule," The tip and inner border of the inner malleolus are generally in close contact with the scaphoid, a bursa intervening; in very severe cases I have seen the inner malleolus in contact with the internal cuneiform, a short flat ligamentous band of great strength passing from one to the other.

In Case 6, this band, besides being attached to the scaphoid, sent fibrous prolongations of considerable strength to the upper and inner surfaces of the internal cuneiform. On cutting through this band close to the malleolus some eversion of the foot became practicable; it was then seen that its deeper part blended with the superior astragalo-scaphoid ligament, and that on continuing to evert the foot this latter ligament became tense, the fibrous expansion representing the inferior calcaneo-scaphoid ligament, and the tendon of the posterior tibial muscle also became tense; the two together effectually prevented straightening of the foot, even when all the other soft structures had been completely removed.

When there is much inversion, the capsular ligaments connecting with each other the scaphoid, the internal cuneiform, and the base of the metatarsal bone of the great toe, all assist in maintaining the deformity. The plantar ligaments—especially the long and short calcaneo-cuboid—are likewise short, in cases in which the original deformity has been increased by walking on the foot; and these ligaments, by reason of their great strength, play a very important part in maintaining the deformity. The above-mentioned ligaments are chiefly concerned in the varus portion of the deformity. As regards the equinus part, it may be said that the posterior ligament of the ankle-joint is shortened, so that in severe cases, after section of the tendo Achillis, the foot can only be partially extended. Many authors refer to this inability to extend the foot after section of the tendo Achillis, but the majority ascribe it to adhesions within its sheath. In point of fact, there is adaptive shortness of the ligamentous structures of the hinder part of the ankle-joint—chiefly the posterior ligament; and in addition there may occasionally exist fibrous bands within the joint, such as those described in Case 6; at this early period, these bands, it is true, were very delicate, but they would naturally become stronger; so that in children of three or four years, they, alone, would offer very considerable resistance to the readjustment of the foot.

The plantar fascia likewise, "the densest of all the fibrous membranes," is of great strength, and when contracted, in adaptation to the foreshortened talipedic foot, becomes one of the most powerful aids in maintaining the deformed position.

THE MUSCLES.—With the exception of those of the calf represented by the tendo Achillis, I believe the muscles as compared with the ligaments play a very subordinate part in congenital club-foot. Very much more extensile than the ligaments and very much longer, they would yield to mechanical means without tenotomy provided the ligaments did not prevent them.

In several dissections, after section of the astragalo-scaphoid ligament just described, I found that the feet could be put almost straight without any undue force being used while all the muscles remained *in situ*, and that only when the foot had assumed its normal position did some of the tendons become at all tense; especially to be remarked were the tendons of the tibialis anticus and of the extensor proprius hallucis muscles; in a less degree also the tibialis posticus muscle. The extensor longus digitorum, on the contrary, began to relax as the foot was straightened, and was almost flaccid when the foot was in the normal position. A glance at a recently dissected talipedic foot will easily explain this; it will show that this relaxation of the digital extensor will be in direct ratio to the amount of inversion which has been overcome, for the muscle is tense only when the foot is inverted. The tendo Achillis, alike from its size and position, plays an important part in maintaining the faulty position of a congenital talipes. I have never seen a case in which this muscle was not shortened. In some cases the shortening is so marked that an interval of two inches between the cut extremities would occur, if the foot were replaced in its normal position. In talipedic feet this tendon will be found attached quite close to the internal border of the posterior surface of the os calcis, a fact which itself will quite account for some of the twisting inwards observed in this bone. Next in importance come the muscles of the sole of the foot, as much by reason of their short length, and therefore their limited capability of being elongated, as by their position; chief among these is the abductor hallucis muscle.

THE BONES.—I have already alluded to the condition of the tarsal bones; the observations were made on fœtal or infantile bones, or in those of quite young children, with the object of discovering primitive deformities as distinguished from secondary ones, such as may have resulted either from walking on the unreduced talipedic foot, or as might be the result of mechanical instrumentation used to reduce the deformities.

Astragalus.—A study of the cases related in Chapter I will show that in one well-marked case there was no appreciable change in the conformation of the astragalus, from which circumstance it must be presumed that abnormal conformation is not the cause, that it is not even an essential element in club-foot. Nevertheless, the neck of the astragalus does present greater obliquity than is normal. The conformation commonly met with may be described as an approximation to that of the Simian type, that is to say, the neck of the bone inclines inwards, forwards, and downwards with greater obliquity than in the normal foot. The average obliquity in the normal infantile bone is $35°$, as against $10·65°$ in the human adult; in the talipedic (infantile) astragalus the mean obliquity of the neck measures $49·6°$. In an adult chimpanzee this angle measures $27°$; in a young adult orang it measures $45°$. Further, the joint surfaces are altered and misplaced, being increased in extent in some directions, and redundant in others, evidently pointing to the fact that they are developmentally laid down on normal lines, and have been altered in response to causes acting subsequently. These points have, however, already been discussed.

Calcaneum.—The chief deviation in the shape of this bone consists in an incurvation of the anterior extremity, so that its outer surface comes to form the arc of a circle : its cuboidal surface, instead of looking directly forwards, looks inwards and forwards. This incurvation is produced by the traction, exerted by the inverted foot on the dorsal calcaneo-cuboid ligament.

All the other bones of the foot, though not materially altered in shape, have an inclination inwards, as is shown in the horizontal section of the foot (Fig. 37).

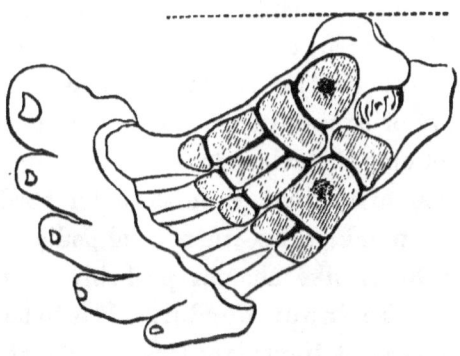

Fig. 37.—Horizontal section through a talipedic foot, showing general incurvation of the tarsal bones.

It is quite easy to understand why the treatment of such a deformity should commence early in life before ossification has taken place; and it will be evident how a curved calcaneum, when ossified, will interfere, with all the power that an arch possesses, with the restoration of the foot.

Of hardly less importance is the altered condition and shape of the articular surfaces of the tarsal bones. In Case 1 (page 10), I found that the trochlear surface of the astragalus was extended backwards, and proportionately lessened in front. The navicular surface was also altered in shape; in both instances the unopposed (redundant) portions were covered with loose connective tissue. A corresponding condition existed on the tibio-fibular surface, and on the navicular bone. Unless these joint-surfaces are early brought into use, as by placing the foot in its normal position, and practising passive movements, so as to develop their synovial covering, the difficulty in effecting complete recovery will be very great.

THE SKIN.—Lastly, as a material part of the anatomy of club-foot, must be mentioned the skin. For though not appreciably altered, the skin becomes a considerable hindrance to the restoration of the foot in severe cases of equino-varus. This hindrance may be gauged by noticing the extent to which the supplementary skin on the convexity, that is, on the outer border, of the foot and ankle wrinkles up when a talipedic foot is put straight; for by a like amount probably is the skin deficient on the inner border. It is to be regretted that we cannot as yet transplant this redundant skin. A great deal of stretching is required before the tendency to revert back—the elasticity of the skin, in other words—is overcome. In some few cases it is advantageous to incise the skin in two or three places; but this should only be done when the deep wounds made by the tenotomes in dividing the tendons and ligaments have quite cicatrised.

UNTREATED AND RELAPSED FORMS OF TALIPES.—I have given the anatomical varieties of club-foot and the conditions, for the most part, as seen in infants and young children. In later life the congenital deformities become greatly exaggerated, and complications arise. These latter, however, must not be confounded with the radical changes present at birth, as they are no essential part of the deformity. Reference has already been made to the atrophic condition of the muscles of the talipedic leg. Not only do the muscles not develop, but they atrophy and undergo such degenerative changes that treatment fails to recover them. Hence success in the treatment of talipes depends very much on the period at which it is commenced; if undertaken late in life, whether for the first time or in relapsed cases, it can never be so full and complete as when undertaken and carefully followed out during early life.

The bony framework of the foot, on becoming fully ossified, no longer lends itself to rectification. More especially the obliquity of the neck of the astragalus and the incurvation of the calcaneum become permanent and most effectually oppose reduction of the deformity. The joint-surfaces, also, become confirmed in their abnormal shape, and the unopposed portions less and less suitable for arthrodial purposes. The ligaments of the misplaced tarsal joints become set, and impede still further the surgeon's efforts to restore the foot to a normal condition. The whole limb indeed grows more slowly than its fellow, so that some compensation will be required, in an extra thick sole to the walking boot.

It must be conceded that the subjects of both single and double congenital talipes manage to get over a good deal of ground, and even perform heavy labour, with but little apparent trouble to themselves. It is this fact which makes me very chary about interfering with old cases of the congenital deformity, say after twelve or fourteen years of age. But circumstances alter cases. An improved appearance of the foot may be of more value than utility alone to a person of independent means, whereas a labouring man will be served by his old misshapen foot better than by one rectified in shape by tarsectomy, but on which he cannot walk without inconvenience, and which requires expensive apparatus for the maintenance of its new position afterwards.

Fortunately these cases will become fewer and fewer,

as more rational methods of treatment are introduced, and as the importance of early treatment becomes more widely known and appreciated.

Thus, it will be seen how largely club-foot is a **condition, in which every structure entering into the formation of the foot participates;** it will be obvious that the earlier the deformity commences the more radical this participation will be. Our treatment must be guided by this knowledge; to cut long slim tendons such as those of the anterior and posterior tibial muscles, while structures such as the tarsal ligaments are disregarded, is, surely, a mistake, and is almost certain to lead to disappointment.

CHAPTER IV.

THE TREATMENT OF CONGENITAL CLUB-FOOT.

PART I. THE SUBCUTANEOUS DIVISION OF TENDONS AND LIGAMENTS.

Objects of tenotomy and syndesmotomy—Indications—Process of repair in tendons and in ligaments—To what extent should deformity be rectified after operation ?—Rules for division of tendons—Rules for division of ligaments.

PART II. TREATMENT OF CLUB-FOOT.

I. *General remarks on treatment—At what age to commence Treatment—Manipulation—Sea-water douching—Galvanism—Flexible splints, Scarpa shoes—Plaster of Paris—Accidents—Tenotomes.*
II. *Treatment of special forms—Talipes calcaneus and calcaneo-valgus—T. equinus and equino-varus. Irregular forms—After-treatment of talipes—Prognosis—Tarsectomy—Conclusion.*

PART I. ON THE SUBCUTANEOUS DIVISION OF TENDONS AND LIGAMENTS.

SECTION of tendons (TENOTOMY) and of ligaments (SYNDESMOTOMY) is practised (1) for the purpose of allowing the talipedic foot to assume a normal position ; (2) to secure a sufficient lengthening of certain of these structures which have been developed too short; and (3) to abolish that constant and unnatural traction on some of the muscles which the misplaced foot, when used for walking, never ceases to exercise, and which, unless counteracted early, invariably results in atrophy and contracture.

Division of ligaments is indicated (1) in a certain number of originally severe cases; (2) in some which have not been treated in early life; and (3) in some relapsed cases. In many cases it will be found that the position of the foot cannot be rectified even after section of several tendons, including the tendo Achillis. This is usually attributed to adhesions which have formed between the tendons and their sheaths, sometimes to inefficient operations, sometimes to other causes. In the preceding chapters I have discussed the anatomy of this condition, and shown that it largely depends on the ligaments, which, like some of the tendons, have been developed too short. The extreme shortness and the unyielding nature of the ligaments render it almost impossible to lengthen them by any amount of stretching which can be tolerated by the living foot.

Hence the explanation, I believe, why so many cases of talipes, which have been merely tenotomised, relapse, as soon as they pass from the surgeon's observation.

The process of repair.—Tenotomy and the subsequent repair of cut tendons have been most carefully studied, by no one more carefully than by Mr. Adams. The subject will be found fully discussed in his well-known Prize Essay to which I have already frequently alluded, and to which for further details I would refer the reader. A brief summary of the process is all I shall attempt in this place. I must add something, however, on the repair of ligaments, though the process does not materially differ from that of tendon repair.

The tendo Achillis has no sheath proper beyond the soft cellular tissue and fat among which it lies. On section of the tendon, the proximal separates from the distal extremity for a variable distance, but the two remain connected together by this loose tissue, into the meshes of which more or less blood is effused at the time of the operation. Within a few days the proper reparative process begins. This consists in the production of embryonic tissue in the connective tissue between and around the divided ends of the tendon as well as in that between the fasciculi of the tendon. In cases in which the ends of the tendons are in apposition, or only slightly separated, the new tissue forms a fusiform band of union; should the ends be very widely separated the new tissue will be found in greatest amount around these, and in lesser amount between them, the band in such cases being constricted in the middle of the interval, so as to be hour-glass shaped.

At about the end of a week, the ends of the tendon will be found slightly bulbous owing to this infiltration, and a narrower cord of new tissue, the union even at this period united by being of considerable strength. Most probably the tendon cells take part in the production of the new tissue, as do also the corpuscles of the connective tissue between and around the ends of the tendon. As commonly described, the formation of permanent fibrous tissue is produced by fibrillation of the intercellular substance, the corpuscles persisting as connective-tissue cells. The tissue of the cicatrix remains distinguishable from the original structure of the tendon by the fact that the new fibres are wanting in the parallel arrangement, which produces the nacreous or pearly appearance characteristic of ordinary tendon.

The process of repair after section of ligaments differs little, *mutatis mutandis*, from that just recorded in the case of tendons. There is, of course, the additional fact that a joint is opened, but this need not complicate matters, provided it is done subcutaneously. After section the edges of the incised ligament are separated by any passive movement which may be applied to the foot, and a variable amount of blood is effused into the interval so formed and into the surrounding cellular tissue. Subsequently embryonic tissue is formed, the cut surfaces of the ligament participating in the process, just as do the cut extremities of the tendons. This embryonic tissue gradually becomes organised into fibrous tissue; and the edges of the ligament are again united by means of this intermediary tissue, while the integrity of the joint is re-established. Some very slight adhesions may occur within the joint, and in view of this possibility, especially in the case of the ankle-joint, it is well to commence passive movements quite early, say at the end of a week after the operation, by which time the external incision will have firmly healed.

After division of a tendon, but little separation takes place between the divided extremities, provided the foot be left to itself. The amount of this separation is only appreciable during life in the tendo Achillis; the tendons of the tibial and other muscles are so small and so hidden beneath the subcutaneous fatty tissue that any interval which may take place between the tenotomised extremities cannot be recognised by the finger. Even in the tendo Achillis, however, the amount of retraction which spontaneously takes place is very small indeed, especially when the tenotomy is performed under chloroform. Should any attempt be made to place the foot in its normal position, in a case of talipes equinus for instance, a considerable gap will be formed, varying from half an inch to two or even three inches. The amount of this separation depends therefore upon the amount of rectification of the foot, which the surgeon accomplishes, rather than upon any inherent tendency in the muscle to contract.

One of the most interesting questions for discussion in connection with tenotomy is, How far should the ends of the tendon be separated at the time of the operation? Is it better, in other words, to place the foot back in its deformed position and wait for some days before attempting rectification? Or should the foot be put as straight as possible immediately after tenotomy? Led on by my own observation and by the experience of other workers, for many years past, I have been in the habit of rectifying the foot as fully as possible immediately after operation, and I have never had reason to regret this, nor do I know of any recorded cases where injury has resulted from the practice.

With a view, however, to elucidate this question practically, I recently undertook some observations on dogs with the co-operation of my friend, Mr. VICTOR HORSLEY, whose kindness I would take this opportunity of acknowledging. On July 18th, under ether, the tendo Achillis of a medium-sized dog was divided, the same care, the same method, and the same precautions being adopted as in the human subject; a collodion and gauze dressing was applied, the foot being fully flexed in order to separate as widely as possible the divided ends of the tendon; in this particular instance the interval measured two inches. On July 26th, it was noted that a band of union could be distinctly felt between the divided ends, which were both, especially the proximal one, slightly bulbous. On July 29th, the dog was springing about without any sign of pain or discomfort. On August 5th, the connecting band was stronger and harder, the dog could extend the limb well, but he walked upon the heel on account of the inordinate length of the tendo Achillis, and not on account of loss of power in the calf-muscles. Varying the conditions a little as in clinical practice, a similar result was obtained in several other cases, both in dogs and rabbits; in none was there any delay in the reparative process, although, in each case, the maximum separation between the cut extremities was brought about at the time of the operation.

It is stated by authors that imperfect union may result, and that sometimes even non-union of the divided tendons is possible. I have never seen such an occurrence; on *a priori* grounds I should incline to think that the constant interference with the process of repair after tenotomy, which must result from early commenced and frequently repeated attempts to elongate the newly-formed tissue, is more likely to give rise to such an untoward accident than any other plan that could be adopted. On the other hand, nothing favours repair so completely as rest. For these reasons I am inclined to recommend that the foot be adjusted immediately after operation and be then left at rest until repair of the divided structures has taken place.

In some cases of acquired talipes, and chiefly in the paralytic forms, where the reparative power and general vitality are reduced greatly below par, the foot after division of the tendon becomes so flail-like that it can be extended or flexed much beyond the normal limit; it is very necessary in such cases to carefully fix the foot well within the normal limit, or some loss of power may result.

RULES FOR THE SUBCUTANEOUS DIVISION OF TENDONS.

Tendo Achillis.—This tendon spreads out somewhat at its lower end, so that the narrowest part is about one inch and a half above its insertion; it is quite superficial and is separated from the deeper structures by loose connective and adipose tissues. Small branches of the posterior tibial and of the peroneal arteries are found just at this point, and one or other may possibly be cut at the operation.

The patient should lie on the face; an assistant then takes hold of the foot and puts the tendon on the stretch by flexing the ankle-joint; it may be cut either from the inner or outer side as most convenient to the operator, and either from below upwards or from above downwards. After a minute incision has been made in the skin perpendicularly to the surface close to the tendon's edge, a blunt-pointed tenotome is introduced, and made to pass flatwise beneath the tendon to the opposite edge; the edge of the blade is then turned towards the tendon, and by a slight to-and-fro movement is made to divide it. A distinct snap will be heard, and the assistant, waiting for this, must relax the foot immediately afterwards.

Tendon of the posterior tibial muscle.—At about one inch above the internal malleolus, the spot usually chosen for the division of this tendon, it is in close relation with that of the flexor longus digitorum, and, in a small infant, between those and the posterior tibial vessels and nerve and the flexor longus hallucis the interval is small indeed. In the talipedic foot, the insertion of this tendon is a very wide one; it will be found as a broad fibrous expansion blending with the astragalo-scaphoid, and the anterior part of the deltoid ligaments, not infrequently also with an expansion from the anterior tibial tendon. Normally, its chief insertion is into the tuberosity of the scaphoid; though it is said to be "inserted into all the bones of the tarsus except the astragalus, and into all the bones of the metatarsus except the first and fifth."

ADAMS, following LITTLE, advises that the puncture should be made above the inner malleolus, "just on the turn of the bone, where, in a severe cases, the posterior tibial tendon is slightly prominent. There is no posterior edge to the tibia as in a perfectly ossified bone, and therefore the exact point must be guessed. The altered direction of the tendon, which bears a constant relation to the severity of the deformity, must be accurately kept in view." . . . "The scalpel must be thrust straight down to the tendon, and by a movement of the point, an incision made in the sheath close to the bone. This being accomplished, the scalpel is withdrawn, and a blunt-pointed knife inserted, care being taken that the point passes between the tendon and the bone. If this is done, the knife will be locked so that it cannot be moved from side to side, but if this sensation be not distinct, the tendon has been missed. The knife should then be partially withdrawn, and used gently as a probe. In this way the tendon may be discovered. . . ." The reader, I am sure, will agree with Mr. ADAMS that this operation "is one of great difficulty."

This tendon may also be divided, as pointed out originally, I believe, by VELPEAU and subsequently by SYME, " a little below and anterior to the tip of the internal malleolus." Here it will be found as a broad fibrous expansion, as has already been described; for anatomical reasons I very much prefer this position. I use a somewhat curved tenotome, enter it over the astragalo-scaphoid joint, and make a curved sweeping cut down to the bone, and include (probably) the tendon of the anterior tibial, as also the astragalo-scaphoid capsule (to which further reference will be made shortly).

Tendon of long flexor muscle of the toes.— This is usually divided at the same time as the posterior tibial, if found necessary. ADAMS says, " When the knife is behind the posterior tibial tendon, it may be pushed a little deeper, with the object of including the tendon of the flexor longus digitorum."

If this muscle, or the long flexor of the great toe, offers any hindrance to the restitution of the foot, it will be best appreciated in the sole of the foot, just beneath the astragalo-scaphoid articulation, and I believe the tendons can be most easily divided in this place. By passing the curved tenotome somewhat deeper and close to the bone, as in division of the posterior tibial tendon described in the foregoing paragraph, the tendons can be simultaneously divided just where they cross each other.

Tendon of the anterior tibial muscle.—This tendon passes across the anterior border (crest) of the tibia at its lower end to be inserted into the inner and under surface of the internal cuneiform bone and base of the metatarsal bone of the great toe; it crosses the inner portion of the ankle-joint, and can be most easily felt and reached just external to, and in front of, the anterior border of the internal malleolus. The *dorsalis pedis* artery is well out of reach.

Tendon of the extensor (proper) of the great toe.—This tendon, when contracted and likely to require division, stands out so prominently that there is no difficulty. The most convenient point is just over the metatarso-phalangeal joint. The tenotome is entered at one side of the tendon, and then made to pass beneath it; the tendon is divided towards the skin, the tip of a finger assisting to steady the tendon, and press it towards the blade.

Peroneal muscles.—The tendons of these muscles may conceivably require division. They will easily be made out just above and behind the external malleolus. The tenotome should be entered between the tendons and the bone. The long and the short muscle will be divided simultaneously.

Abductor muscle of the great toe.—This muscle can most advantageously be divided near its origin from the inner tuberosity of the os calcis. It is very closely identified with the plantar fascia, and the two are often divided simultaneously. A blunt-pointed tenotome should be introduced on the flat just in front of the tuberosity between the skin and the deeper structures; the cutting edge being then turned towards the bone, all the structures may be divided until the bone is reached, the foot being firmly redressed while the section is being done; any few uncut fibres can easily be ruptured by suitable manipulation. Some further elongation can be obtained, if required, by cutting this muscle near the base of the metatarsal bone.

Plantar fascia.—Either instead of or in addition to the operation just recorded, the plantar fascia may be divided at any point in the sole of the foot, where resisting bands can be felt. This is a most unyielding structure, and more can be accomplished in a few minutes by the judicious use of a tenotome than from the application of apparatus during months. It is a structure which claims great attention in all cases of congenital club-foot.

Rules for the Subcutaneous Division of Ligaments.

Anterior ligament of the ankle.—Section of this ligament, which is very thin, is rarely required; it is most easily effected by entering the knife at the anterior border of the inner malleolus, and passing it horizontally into and across the ankle-joint. In this manner the anterior border of the internal lateral ligament, which is always very strong, will be divided simultaneously with the anterior ligament proper. The foot should be forcibly extended, and the section accomplished by one or two touches with the point of the tenotome, so as not to wound any other structures in front of the joint.

Posterior ligament of the ankle.—A double-edged spear-shaped tenotome with a rounded shank may be passed vertically through the centre of the tendo Achillis a little below the spot where tenotomy is usually practised; the blade is then turned flatways and made to enter the capsule about its centre. The ligament is now put on the stretch by flexing the ankle, and is then cut as much as may be necessary, first on one side of the median line, then on the other. The advantage of a double-edged knife is obvious; it does not require to be turned, and there is thus less disturbance of the soft areolar tissue around the joint. A snap less marked but somewhat similar to that felt after section of the Achilles tendon will be experienced when the ligament yields. After the chief part of the ligament has been divided, any remaining fibres can generally be snapped by suitable manipulation.

In practising these operations in the dead-house, I have found it difficult to completely divide the ankle-joint ligaments; they merge imperceptibly one into the other; the surfaces to which they are attached are very uneven and irregular, and so interfere with the course of the blade. In attempting to divide the posterior ligament I found, on dissection, that the central portions only had been divided; nevertheless, flexion of the joint was greatly facilitated, and the remaining fibres were broken down without any great difficulty. In fact, in all cases of syndesmotomy, forcible manipulation should be practised, both during and after the operation, as much for the purpose of making the ligaments tense and more easy to cut, as with the intention of breaking down any fibrous bands which may have escaped the knife.

The foot represented in the Plate is in a condition of extreme equino-varus, as indicated in the little outline plan appended. The arrows point to corresponding parts, and show the direction in which the incision into the astragalo-scaphoid capsule ought to be made.

DESCRIPTION OF PLATE.

1. Musc. tibialis anticus.
2. ,, extens. proprius hallucis.
3. ,, ,, longus digitorum.
4. ,, ,, brevis digitorum.
5. ,, abductor hallucis.
6. ,, flexor longus hallucis.
7. Inner tuberosity, os calcis. Tendo Achillis.
8. Navicular bone, in contact with inner malleolus.
9. Internal cuneiform bone. Small openings have been made into the capsular ligaments which connect this bone with the navicular, and with the base of the first metatarsal bone.
10. Musc. tibialis posticus.
11. ,, flexor longus digitorum.
12. ,, ,, hallucis.
13. Posterior tibial vessels and nerve.

NAT. SIZE

S.G Shattock ad nat. del

Darielsson & Co, lith

"**Astragalo-scaphoid capsule.**"—This dense structure is fortunately quite subcutaneous, and so placed that it can be divided without risk to any other structure. The tenotome, being held vertically, with the edge forwards, should be entered immediately in front of the anterior border of the internal malleolus, the blade being kept as far as possible between the structures to be divided and the superjacent skin. In the next stage, the blade is turned towards the surface of the ligaments, and, by means of a gently-sawing motion, is made to divide it.

As the superficial fibres are divided, deeper ones come into play, and must in their turn also be cut until the tarsal bones are reached. By keeping the knife close to the bones and directing its point to the plantar aspect of the foot the calcaneo-scaphoid part of the ligament can be easily divided. Reference to the Plate will show how the "astragalo-scaphoid capsule" including the tendons in its immediate neighbourhood can all be divided at one time.

Should this prove insufficient to set the foot free, the tenotome may be again introduced just in front of the inner tuberosity of the calcaneum, and the abductor muscle and plantar fascia divided separately or together.

Long and short plantar (calcaneo-cuboid) ligaments.—These are so intimately blended together that they may be dealt with as a single ligament. The tenotome must be entered as nearly as possible over the calcaneo-cuboid articulation on the outer border of the foot; if there is much inversion this point will appear to be on the sole of the foot, and allowance must be made for the altered position. The blade should be kept close to the bone, and be made to follow the direction inwards across the sole which this articulation takes (*vide* Fig. 42, page 98); in this manner the two ligaments will be divided at once, traction being made on the forepart of the foot, so as to make the ligaments as tense as possible.

In severe cases of talipes, where the inner border of the foot is much incurved, the ligaments between the internal cuneiform and the navicular, and those between the internal cuneiform and the first metatarsal bone, may require division. The tenotome must be introduced over the respective joints and the ligamentous structures, more or less capsular ligaments, divided as in the preceding cases.

Part II. The Treatment of Club-foot.

I. General Remarks.

At what age should treatment be commenced?—Whatever be the variety, and whether a slight or a severe case of club-foot, it may be laid down almost as a canon that treatment should be commenced immediately after birth.

From what has been said it will have been gathered that all the structures of the foot participate in the deformity; a glance at Fig. 37, page 67, will show how important it is that treatment should commence early, before the cartilaginous basis of the tarsal bones and their misshapen articular surfaces have had time to harden or ossify. It will hardly be necessary to add that all manipulations must be in keeping with the delicate nature of the affected structures at this early period of life. One of the further results of leaving talipes untreated is slow and imperfect development of the muscles of the leg from disuse, and this is not easily made up for if treatment is long delayed.

Treatment must have a two-fold object.—It must (1) not only overcome the anatomical deformity, and replace the foot in its normal position, but must also (2) retain it in this newly-acquired position until the physiological development of the foot renders such mechanical aid unnecessary. For a long time mechanical measures were solely relied upon. It was Stromeyer's introduction of subcutaneous tenotomy which showed the advantage of this over mechanical means, and it is the accumulated experience of all surgeons since that time which now recognises that neither the one nor the other alone suffices, and that it is from a combination of the two that the most satisfactory results may be obtained. Before detailing the treatment of special forms of talipes it may be well to refer to such matters as concern all forms.

Manipulation.—Manipulation is a very necessary and a very important part of the treatment of every case of talipes, whether mild or severe. In infants, a great part of the deformity can be overcome by gentle and well-applied manipulation. Besides the elongation of all the shortened structures that can be accomplished by its means the general nutrition of the limb is kept up, and the atrophy which would certainly follow the but very limited nature of the movements of which these talipedic limbs are capable, is minimised, not to say prevented, by gentle and persistent manipulation and shampooing. The limbs may be sluiced with tepid sea-water, or with water in which sea-salt has been dissolved; then the muscles should be kneaded with the tips of the fingers, and the limbs generally well rubbed. The deformed foot should subsequently be "worked," that is, by a series of short, frequently repeated movements it should be placed in or directed towards its normal position. At first, on releasing the foot, the old position will be reassumed, but after a while this tendency will be less and less marked. Professor SAYRE truly says, "Manipulation may be regarded as the natural remedial agent for the cure of a deformity . . . . There is an intelligent touch (in the human hand) that admonishes you of the amount of resistance present, the amount of force required to overcome it, and when you should stop its exercise." The *rationale* of manipulation is to be found in the increased circulation throughout the limb, and in the improved nutrition which is thus secured.

Galvanism also will prove useful in cases where much

atrophy has taken place; applied gently and with skill it can rarely do harm ; when, either from the nature of the case, or from accidental circumstances, treatment has been postponed, or is likely to last for any length of time, galvanism is also indicated, and will be advantageous. I do not think it exerts any special influence ; it is merely a further means of keeping up the circulation, and of setting the muscle-fibres in action. An efficient mode of application is to place the foot in a basin of warm salt water along with one rheophore, and with the other rheophore, terminating in a large soft sponge, to stroke the limb, beginning on the buttock or thigh, and working down towards the foot; the application should last from five to fifteen minutes, and the current should be nicely regulated to the personal susceptibility of the patient.

Flexible splints.—In the milder cases, and in young infants, after manipulation, &c., light flexible splints may be applied. These splints, made of tin plate and covered with wash leather, may readily be bent into any required shape. Thus, if the foot cannot be got straight, the splint must be adapted accordingly, and altered from day to day as improvement goes on. In young infants, whose skin easily galls under pressure, great care must be taken. A little absorbent cotton-wool, with an abundance of dusting powder over the bony prominences must be used to protect the foot as much as possible ; open wove (absorbent) bandages are the most appropriate, being softer than ordinary calico, and cooler than flannel.

Scarpa's shoes.—While admitting the great value of the mechanical contrivances known under the generic name of Scarpa's shoes, I must confess that I rarely use them except in old relapsed cases, or in cases which first come under observation some years after birth, or in the severest forms of the congenital deformity. Among the advantages of these shoes, which can be easily removed and readjusted—two or three times every day if necessary—may be mentioned that the foot and leg can be shampooed or galvanised, and that manipulation by the hand can be systematically carried out. To be effectual, however, they must be applied by, and remain under the daily control of, the surgeon. Even then the greatest care is necessary to secure the advantages without blistering the feet at the points where pressure is being exercised. The shoes invented and used by Mr. ADAMS are among the most ingenious and serviceable I am acquainted with. For old and relapsed cases nothing can be better; if persevered with long enough they are capable of reducing very severe forms of deformity.

Plaster of Paris.—The applicability of this substance has greatly increased since the use of muslin bandages impregnated with the plaster has been in vogue. Plaster bandages, moreover, are inexpensive, easily transported, and can be obtained almost anywhere. The coarsest muslin should be selected, and, having been cut into the desired lengths and widths, plaster should be rubbed in until the meshes of the muslin are quite full; the bandages should not be rolled up too tightly, so that they may rapidly and completely soak through, when put into water just before use.

I advocate and use these bandages in preference to a Scarpa's shoe for the reason that a more widely diffused pressure can be brought to bear on the deformed foot. In a Scarpa's shoe the pressure is much more localised; the straps for the most part are very narrow, and take their bearing on some fixed and uniform spot—hence the liability to pressure sores. On the other hand, a plaster bandage can be moulded to the foot, and the pressure made to bear, not on points, but on areas of the foot;

MODE OF APPLYING PLASTER BANDAGES.

thus they can be worn with greater comfort and without fear of producing sores. They are messy to put on, and by reason of their hardness are troublesome to change—but these are trifling inconveniences in face of their other manifest advantages. I have elsewhere given reasons why I prefer to utilise to the full and immediately any advantage which is gained by surgical operation, or by forced manipulation; to secure permanently this advantage, I am sure there is nothing better than a carefully-adapted plaster bandage.

A soft bandage from the toes to the knee should be applied before the plaster bandages are put on. This is better than cotton-wool, for the pressure is more equal, and there is less fear of sores. Plaster bandages may be allowed to remain on from a week to a month, according to circumstances; if gradual extension is being practised, the bandages will require changing frequently, a little gain being effected at each fresh application. But if the foot has assumed a normal position, the bandage being required merely to maintain that position, then one month's interval between the changes is not too long. The surgeon must adapt himself to circumstances; a change of bandage gives him the opportunity of actually seeing how the foot is going on, and allows of any manipulations which may be considered desirable. In growing infants and children a more frequent change is desirable than in adults, for then no sort of interference with the growth of the foot can result. Besides this, orderly movements of extension and flexion help to restore the joint surfaces to their normal limits and conditions. For hospital patients, plaster bandages are a great boon as they save much expenditure in costly apparatus.

Accidents.—In a few cases inflammation and suppuration have followed an operation. It is now generally admitted that such mishaps depend on the use of dirty instruments, sponges, or fingers, and that they are manifestations of some infective process. Occasionally also erysipelas has followed. It is needless to point out that the most scrupulous cleanliness is as imperative in this as in all other surgical operations. If an artery should be accidentally wounded, troublesome hæmorrhage would occur. Such an accident might or might not be recognised at the time, or it might only be found out on the development of hæmorrhage, or of a traumatic aneurism. If suspected at the time, it would be better to completely divide a wounded artery, and then apply a graduated compress, and a bandage; if only found out subsequently, the accident will have to be treated on general principles. It is not improbable that the posterior tibial artery is often cut, and that no ill-consequences follow, so free and full are the anastomoses in the foot.

Tenotomes.—In the repair of tendons and ligaments, success depends largely on their division being accomplished subcutaneously. For this purpose the incision in the skin, through which they are reached, should be as small as possible. A sharp-pointed tenotome is used for the purpose, and should be passed in perpendicularly to the surface. Some surgeons make a "valvular opening," that is, the skin is pushed over the proper spot, in order that when the knife is withdrawn the superficial and deep wounds shall no longer correspond, the skin sliding back to the normal position. On withdrawing the sharp-pointed knife (which is only wanted to make a track through the skin and fascia), the blunt-pointed tenotome is introduced in its place, and carefully directed towards the tendon to be cut; the cutting edge is then turned towards the tendon, which is divided by a gentle cutting movement. Care must be taken not to penetrate more deeply than necessary, nor to pierce the skin opposite, nor to cut the skin at, the seat of entrance, nor to cut the skin over the tendon after the latter is divided; the tip of the middle finger, placed close to the point where the blade enters, should serve as a fulcrum, the cutting being affected largely by the point of the tenotome, to which a slight to-and-fro movement must be given. When in the proximity of an artery or a nerve, it is better, if possible, to enter the knife on the same side, and cut away from them.

If several tendons or ligaments are to be divided at one sitting, I apply an Esmarch's bandage; in this way all hæmorrhage is prevented, as the pads and bandage can be applied before the Esmarch is removed. Too much blood poured out into the sheath of a tendon is apt to interfere with rapid healing, and possibly also with the movements of the tendon. The external incision may be covered with a circlet of strapping, and over this graduated pads of absorbent wool or lint, and a carefully adjusted bandage. I think wool is better than lint because pressure can be more easily regulated, and any tendency to hæmorrhage more thoroughly controlled.

The annexed woodcuts show the knives which I prefer to those ordinarily used. The cutting blades are usually made too long. As a rule they need not exceed half an inch; the shank should be strong, and rounded so as not to cut the skin or unduly enlarge the entrance wound; the shank must vary in length from a quarter to three quarters of an inch. The curved tenotomes must have a cutting blade varying from a quarter to half an inch. A double-edged, spear-headed tenotome will be found very useful for the posterior ligament of the ankle; the shank should measure from one inch to one and a half, the cutting blade being about a quarter of an inch. I do not recommend very slight knives, as they are apt to snap, especially when old tendons and ligaments are being

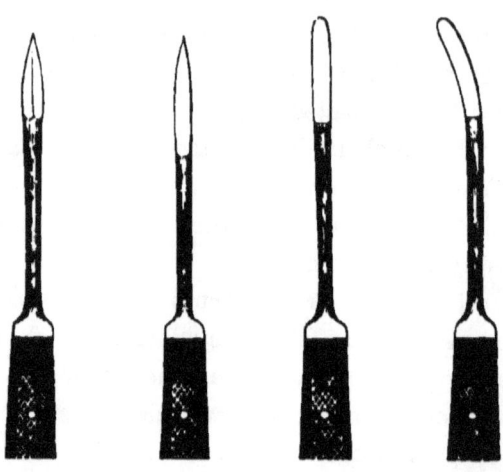

FIGS. 38, 39, 40, 41.—Tenotomes, as recommended by the author.

operated upon, nor do I think that anything is gained by having very small handles—delicacy of manipulation rests with the surgeon, and not in the instruments.

The knives represented in the adjoining woodcuts are made to exact size and shape; the surgeon will, however, do well to be provided with more than one set of different sizes.

II. Treatment of Special Forms.

Talipes calcaneus and calcaneo-valgus.—By reason of their comparatively late onset, and perhaps also in consequence of their anatomical conformation, the treatment of these forms of talipes does not, as a rule, present any difficulties. Manipulation, if commenced very early, will suffice in all but the severest cases. The mother or nurse should be instructed to grasp the foot, and by steady traction endeavour to get the toes down, that is to extend the ankle-joint. This should be done on every possible occasion. If the foot continues to resume the calcaneous position after manipulation, and notwithstanding that the shortened structures appear to be yielding, a flexible splint suitably bent may be applied to the anterior aspect of the limb from the toes up to the knee; in this manner the position of extension will be maintained. The splint should be removed two or three times a day, or oftener, in order to manipulate the joint.

In a few cases, 2 or 3 per cent., division of the structures in front of the joint may be necessary. The anterior ligament is a thin membrane, but it is tough. On the other hand, the anterior border of the internal lateral ligament is a very strong band, and materially aids in maintaining the position. The anterior fasciculus of the external lateral ligament is also strong and helps to maintain the everted position; these structures must therefore be divided. Another attempt may then be made to flex the foot; should this also prove unavailing, division of the tendons of the anterior tibial and other muscles may be practised. As soon as a sufficient amount of extension has been attained a plaster bandage should be applied, and the foot fixed; the bandage may be changed every fortnight, three weeks, or once a month according to circumstances. After a few weeks, some movable splint can be substituted for the plaster. For young children a poroplastic shoe will be found as convenient as anything.

The prognosis in calcaneus and its varieties is eminently satisfactory; of all forms of club-foot, they are the easiest to treat, and the least liable to relapse.

Talipes equinus and equino-varus.—Uncomplicated equinus is almost unknown as a congenital deformity.

The case illustrated in Fig. 30 (page 58) is the nearest approach to this condition which has come under my own observation; the other foot was in a position of marked equino-varus. In this case division of the tendo Achillis, with a little manipulation of the foot, sufficed to effect a cure. Both hands manifested the homologous condition, that is to say, the wrist was flexed, the hand somewhat adducted, and the thumb adducted and fixed in the palm.

For the first stage of club-foot (equino-varus), a narrow, flexible and well-padded splint, the lower portion of which has been bent to an obtuse angle, should be applied to the outer surface of the leg, and fixed with a few turns of bandage; the foot, after suitable manipulation, must then be everted, and bandaged to the splint, the interval between the foot and the lower part of the splint being filled with a pad of wool. The splint should be removed, and the manipulation practised two or three times a day, the foot being over-corrected each time. In this manner the bandage and splint fix the foot in its normal position, and as growth takes place the foot gradually retains this position; when the child begins to walk, the weight of the body completes the cure. Sometimes the tendency to inversion remains marked although there is really no anatomical condition calling for surgical treatment. In such cases, a leather boot with "stiffening" along its inner border will help to keep the foot prone; in addition to this a strap fixed about the centre of the outer border of the boot, with a piece of elastic let in, and carried up the leg to a band fastening either just above or below the knee, may be tried if necessary. Or plaster bandages may be used instead; they must be renewed at short intervals.

For the second stage, instead of, or in addition to, the above measures, some surgical operation will be desirable. As a rule, section of the tendo Achillis will suffice. For many years after the invention of subcutaneous tenotomy surgeons contented themselves with this in all cases, and treated any inversion which afterwards remained by mechanical means. In young subjects, division of the Achilles tendon very often suffices even when there is considerable inversion, the latter being treated by manipulation, plaster bandages, or mechanical shoes; the younger the patient, the more hopeful the practice. If, on examination with the finger tip, the anterior portion of the internal lateral ligament be found tense, its division must of course be undertaken.

For the third stage, either instead of or in addition to the measures already alluded to in the preceding paragraphs, something more radical in the way of operation will be required. As I did not attempt any hard-and-fast definition or description of these cases, which differ very much among themselves, so I shall abstain from laying down rigid rules for treating them. It will suffice to say that the experience of former cases of the kind must be brought to bear upon the individual case under discussion, and that, after careful examination, such structures must be divided as are thought or known to be shortened, or instrumental in maintaining the deformity.

I have elsewhere mentioned that in congenital talipes, the tendo Achillis is attached quite close to the internal border of the posterior surface of the os calcis, and that this fact accounts for some of the twisting and inversion of the sole of the foot. It is on this account that I think it well to cut the tendo Achillis quite early on in the treatment of these severe cases; for it is impossible to properly gauge the part played by other structures so long as the Achilles tendon remains uncut. The old argument, to leave it quite to the last, that it may serve as a *point d'appui*, through which to exert mechanical force on the other structures, is wrong in principle since tenotomy has been introduced. When treatment was entirely mechanical, there was some reason in the argument. Now that section of the offending structures is regarded as the chief means of treating these deformities the assistance of the tendo Achillis should no longer be required; if it prove necessary in any given case, it may be taken for granted that the offending structures have not been divided.

As regards the inversion of the foot, of chief importance is the astragalo-scaphoid capsule. With the finger tip, the thickness of this structure, its direction and extent, must be made out, and its division effected in the manner already described. As one set of fibres is cut through, others will come into prominence, and these must be divided until all resistance at this point is removed. The plantar fascia must then be carefully examined, and all tense fibres divided. When the foot still remains more or less rigid, even after complete division of these structures, it will be found to depend on the condition of the long and short plantar ligaments. The adjoining sketch of the sole of a foot, Fig. 42, which was taken from a case recently under my care, shows how the forepart of the foot is approximated to the heel, a deep furrow intervening. It is in this furrow that the offending ligaments must be attacked, and divided in the manner described on page 83. When this is fully accomplished the sole of the foot will unroll, the inversion will yield to gentle traction, and the deformity disappear. It may be necessary to break through a few fibres in order to secure complete restitution of the foot; this may be done without risk or danger.

The operation completed, graduated pads of absorbent (or antiseptic) wool should be placed over the punctures, and a soft bandage applied. Over these a plaster-of-

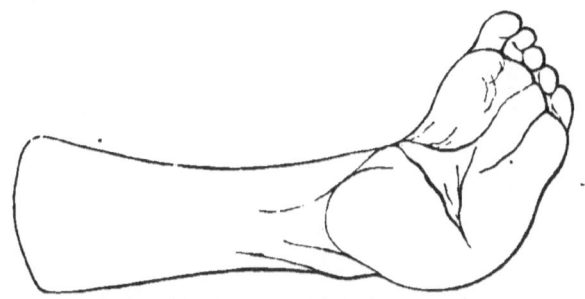

FIG. 42.—Sole of the foot. The transverse line in front of the heel shows approximately the position of the plantar ligaments which may require division.

Paris bandage, the foot being placed and held in as normal a position as possible, until the plaster has set firmly. This bandage may be left for a week or ten days, by which time the punctures will have healed. It may then be removed, and the foot manipulated, by which means perhaps some additional mobility will be gained, before the plaster bandage is renewed.

It must not be supposed that I advocate syndesmotomy for all cases. Quite the contrary. If taken in hand early, a large proportion of the total number of cases of talipes will yield after division of the tendo Achillis. In other cases, time is saved by an operation, although with time much could be accomplished by mechanical means alone. In a certain proportion of cases, however, syndesmotomy is the only operation, which really meets the requirements, as indicated by the anatomy of the deformity.

It will be asked to what extent should this operative

treatment be carried out at one sitting? To this I reply that there is no hard-and-fast rule of practice. The patient being chloroformed, and an Esmarch bandage having been applied, the surgeon can accurately gauge the effect of each fibre he cuts. If the foot comes straight without difficulty, he will have gained his object; if it does not come straight, even with the use of force, then there are still impediments to overcome, and to this task he must devote himself.

After-treatment of talipes.—The deformity having been overcome, how is the normal position to be maintained? This is a problem of hardly less importance than the primary treatment. It may be laid down, that watchful supervision will have to be exercised for many years. The earlier treatment is commenced the better will be the final result; the more complete the removal of those causes which hold the foot in its distorted position, the less risk will there be of relapse. Nevertheless, supervision of the foot must be constant and continuous, in order, on the other hand, that the surgeon may detect the slightest tendency to revert back to its former malposition, and on the other to ensure the fullest possible development of the affected limb. Massage and frictions are among the most obvious means to attain the latter, and must be regularly and systematically practised; while properly constructed mechanical shoes will have to be worn in order to maintain the foot in its newly-acquired position. Physiological development will proceed on normal lines, and little by little at first neutralise and then altogether supplant the old congenital habit.

I have not invented any special instruments, and am not concerned in advocating one form over another; many very ingenious appliances are to be had, and each surgeon will prefer that to which he is accustomed. As DE ST. GERMAIN* says: "*L'homme de l'art tire sa règle de conduite du cas qu'il a sous les yeux.*" It may be stated that the apparatus should, as a rule, reach up above the knee; if there is any tendency in the limb itself to inversion, an apparatus reaching to the pelvis will be required. Cases of crus varum will be a source of considerable trouble, and treatment will extend over a lengthened period. It is important in severe cases to use "night-shoes;" for during sleep, when all volition is suspended, the feet and limbs tend to a position of inversion by their own specific weight, a position which the pressure of the bedclothes rather favours; thus, that which has been gained by day is lost at night. For this reason night-shoes should always be worn; they are made on the same principle as the walking shoes, but are lighter and therefore better adapted for night use.

Prognosis in congenital club-foot.—A favorable prognosis may be confidently given, provided the child can have the necessary attention for a sufficient length of time, and be supplied with such instruments as are required to maintain the foot in the normal position after the operative measures are completed. I have already drawn attention to the great desirability of commencing treatment early—the earlier the better, before the tarsal bones have ossified and become fixed in their abnormal directions. After all resisting structures have been divided there only remains the peculiar shape of the bones, and of the articular surfaces, but these rather incline the foot back to its developmental position. Hence the use of some instrument or other means of fixing the foot in the position it is wished to assume.

* 'Chirurgie Orthopédique,' Paris, 1883, p. 573.

The earlier this new position is assumed the less is the secondary deformity that has to be overcome. For a variable time the foot, when left to itself, will relapse back, more or less, to the deformed position, and little improvement may seem manifest. When the foot begins to assume a normal appearance, and to lose the square-set, somewhat flattened-out appearance so well known in cases of club-foot, it is a sign that growth in the right direction is taking place, and may always be considered of happy augury. Development of the calf of the leg also is another good sign.

As to the length of time necessary for the adequate treatment of any given case, nothing very definite can be laid down; it may be stated generally that the first part of the treatment in a young infant will not be accomplished in less than six weeks as a minimum, and that it may last three months or even longer. By this time all resisting structures will have been cut or relieved in some other way, and the infant may be fitted with an instrument to fix the foot in a straight position. Such an instrument will have to be worn until the child shows signs of walking, and then a boot with side supports will be required. As the child grows, stronger boots will be required. If no relapse has taken place, that is to say, if manipulations have been systematically carried out by an intelligent mother or nurse in the intervals, by the time the child is three years old the deformity will generally have been overcome, and little or nothing of it remain. Something will, of course, depend on the nature and severity of the case, and on the child's health and strength. The older the child when treatment is first commenced, or recommenced, the longer will be the time during which the case must remain under active treatment.

What allowance must be made for possible interference

with the course of treatment by the onset of some of the diseases to which children are liable—whooping-cough, or the exanthemata, or even dentition? When appealed to under such circumstances, I always advise that the talipedic instruments and manipulations should be continued as usual, unless there are very special reasons to the contrary, such as rarely occur.

When the more active treatment is completed, an occasional visit to the surgeon is desirable in order that the instrument may be examined and readjusted to the requirements of the case if necessary.

I am very conscious of the difficulties which surround a case of severe or relapsed talipes, and of the time and patience required to deal adequately with it. The children may become weakly, their limbs wasted and undeveloped, the circulation slow, and the skin unable to resist even mild pressure for any length of time together. Apparatus are costly, and the *res angustæ domi* often such as to add seriously to the difficulties of the case. I have written chiefly concerning club-foot as seen during the earliest months of life; these cases are more amenable to the treatment advocated than are older patients, in whom the tarsal bones may have undergone such alterations in shape as almost preclude a complete cure of the deformity.

For such cases, if it is thought desirable to do anything at all, tarsectomy in one or other of its forms appears the only reasonable treatment to adopt. Mr. DAVIES-COLLEY's operation has given excellent results both at Guy's Hospital, where it was first practised, and at the Westminster Hospital, where Mr. RICHARD DAVY has largely adopted it. In America and on the Continent of Europe also, writers have described operations of various kinds practised for the relief of individual cases. It is, of course, impossible to advocate one in preference to another, as there can be no hard-and-fast line of treatment. Each case must be dealt with on its own merits, and according to its individual requirements. Those who are interested in the subject of tarsectomy may refer, among others, to the writings of DAVIES-COLLEY, DAVY, LUND, PUGHE, KEETLEY in this country; of VOLKMANN, HÜTER, F. BUSCH, BÖCKEL, KRAUSE, and KARL ROSER in Germany, PHELPS and HINGSTON in America; POINSOT in France; and MARGARY in Italy. . . .

I am sanguine enough to belive that there will be fewer failures, fewer relapses, and therefore fewer cases to get into this almost hopeless condition, if the anatomical structures concerned in congenital talipes be thoroughly reconsidered by, and allowed to guide those, who are engaged in the treatment of these deformities.

FINIS.

INDEX.

ADHESIONS, within ankle-joint, 13, 14, 64.
Age at which to commence treatment, 85.
Amniotic fluid, deficiency of, as a cause, 32.
    ,,   adhesions, 42, 43.
Analogous conditions, occurrence of, to talipes, 33.
Anatomical classification of talipes, 53, 54.
Anatomy of talipes, 9—14, 61—70.
Ankle-joint, ligaments of, division of, 81.
Apes, higher, not talipedic, 18, 27.
"Argument, the," 35.
Astragalo-scaphoid capsule, its anatomy, 63.
    ,,   ,,   its division, 83.
Astragalus, of chimpanzee, 17; of ourang, 17.
    ,,   adult, 16; infantile, 16; talipedic, 16.
    ,,   angle of neck of, its measurements, 15.
Atrophy, signs of, generally absent in congenital talipes, 24.

BONES, alterations in shape, as a cause, 25, 26, 27.
  ,,   ,,   ,,   as a consequence, 26.
  ,, long, congenital absence of, 35, 46.
Breech presentations, in relation to talipes, 28.

CALCANEO-VALGUS, talipes, 54, 56.
Calcaneum, conformation of, 13, 66.
Calcaneus, talipes, anatomy of, 13, 33, 55.
    ,,   ,,   treatment of, 81, 87, 92.
    ,,   ,,   comparatively late onset of, 30.
Cases of talipes, dissections of, 6—14.
Causes of talipes, considered, 21 *et seq.*
    ,,   ,,   nerve, 22; bone, 25; mechanical, 31; predisposing, 50; unusual, 35, 36.

"Club-foot" means equino-varus, 58.
Club-hand, 47, 49.
Crus Varum, 61, 62.
　　　,,　　treatment of, 100.

DEGREE of talipes, how to estimate, 62.

ENVIRONMENT as a cause of talipes, 31, 32.
Equino-varus, talipes, 57; treatment of, 94, 95, 96.
Equinus talipes, 57.

GALVANISM, 87.
"Genu Recurvatum," 27, 29.

HEREDITY, 51.

LIGAMENTS, condition of, in talipes, 62, 63.
　　　,,　　division of, in talipes, 72.
　　　,,　　　　,,　　repair after, 74.
　　　,,　　rules for, 81.
Long bones, congenital absence of, as a cause of deformity, 48, 49.
　　　,,　　rotation of, effect of environment, 44.

MANIPULATION and massage, 87.
Muscles, in club-foot, not important, 64.
　　　,,　　tendons of, rules for division, 77.

NERVE causes, considered, 22; refuted, 23, 24.
　　　,,　　lesions, not demonstrated, 23.

ONSET of talipes, date of, 27.
Ovarian tumour, as a cause of talipes, 36.

PLANTAR fascia, 64, 80.
Plaster of Paris, advantages and application of, 89.
Pressure, intra-uterine, evidences of, 42, 45, 49.
Prognosis, in talipes, 100.

RELAPSED cases, 7, 69.
Rectification after operation, discussed, 75.

Rectification, experimental investigation of, 75.
,, in paralytic talipes, caution as to, 76.

SKIN, an important factor in maintaining talipes, 68.
Syndesmotomy, indications, 72; rules for, 81; repair after, 74.

TALIPES, anatomical classification, 53, 54.
,, varieties considered, 55 *et seq.*
Tarsectomy, 7, 69, 101.
Tenotomes, 90, 91.
Tenotomy, indications for, 72.
,, repair after, 72.
,, rules for, 77.
Treatment, after-, 99.
,, general remarks on, 85 *et seq.*
,, special remarks on, 92 *et seq.*
Twin children and club-foot, 45.

VALGUS, talipes, 57.

BY THE SAME AUTHOR.

2nd Edition, pp. 117.

TRACHEOTOMY IN LARYNGEAL DIPHTHERIA:
ITS AFTER-TREATMENT AND COMPLICATIONS.

Some Opinions of the Press.

".... The success attending this mode of treatment has been very satisfactory in Mr. Parker's hands, and those who carefully read his book and weigh his arguments will be impressed with the cogency of his reasoning. Much valuable information on the best ways of dealing with the various difficulties and complications of tracheotomy is contained in the book....."—*Lancet.*

".... The work of a sound and practical surgeon, and one of the best descriptions of the operation and its requirements we are acquainted with. ...."—*Brit. Med. Journ.*

".... In many ways it is a most useful and creditable volume. We are not aware there is another work in the English language devoted entirely to the subject of tracheotomy...... It would be well if the remarks upon early operations were more frequently acted upon...... Taken as a whole, the work is most admirable with regard to its directions as to diet, medicine, local application, and on the nursing and after-treatment of a case of diphtheria and tracheotomy, and shows evidence that the writer has had thorough practical experience of these departments of the subject....."—*The Practitioner.*

"This is an excellent little work, giving in clear and simple language a thoroughly reliable account of the most approved methods of treating laryngeal diphtheria....."—*Bristol Med.-Chirurg. Journal.*

".... His concise and explicit directions will be appreciated by those who have sought in vain in more extensive works information in these particulars....."—*New York Medical Record.*

www.ingramcontent.com/pod-product-compliance
Lightning Source LLC
Chambersburg PA
CBHW030404170426
43202CB00010B/1479